D0713441

in Nontechnical Language

2nd Edition

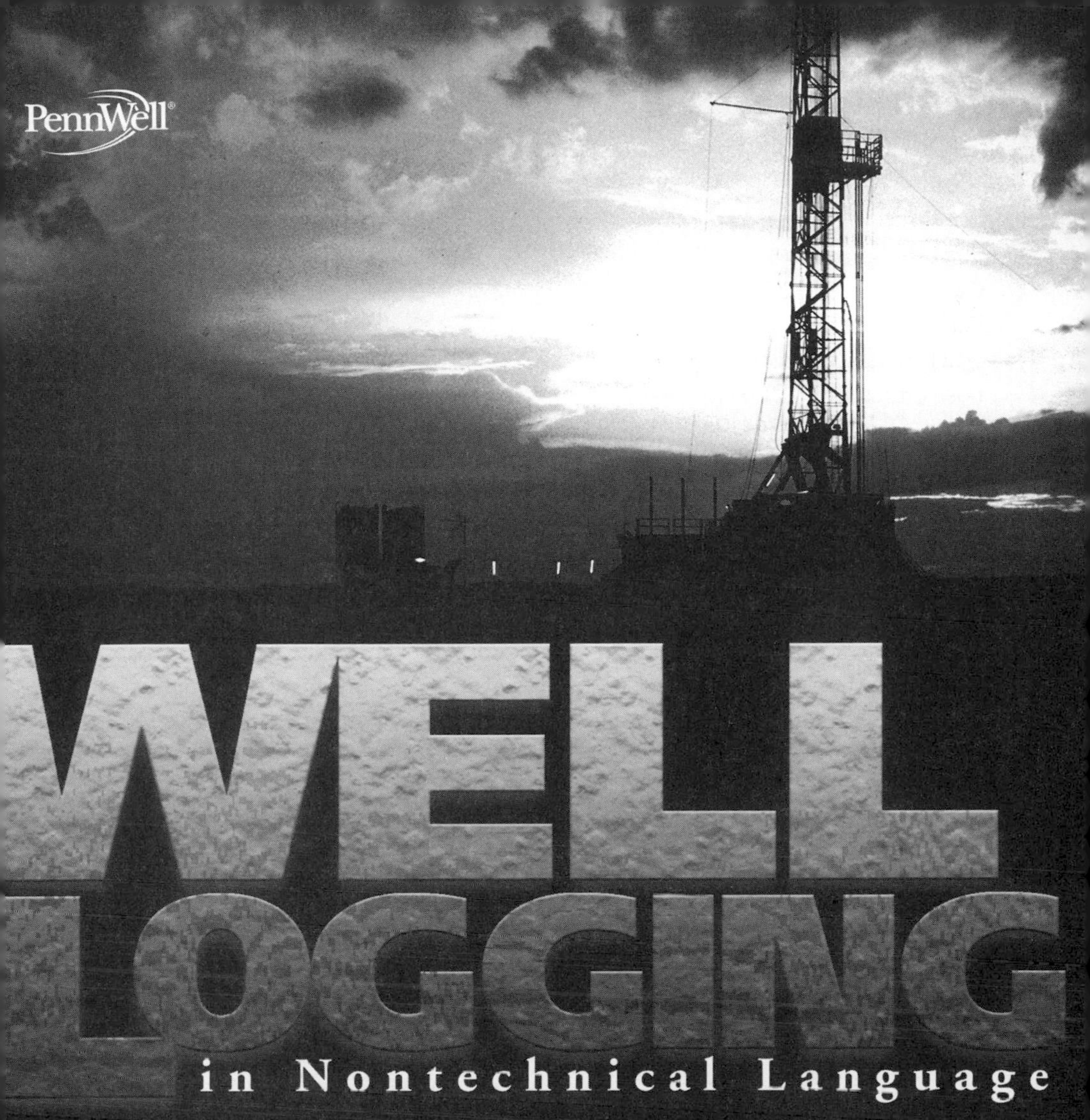

WELL LOGGING

in Nontechnical Language

2nd Edition

David E. Johnson & Kathryne E. Pile

PennWell Publishing Company
1421 South Sheridan
Post Office Box 1260
Tulsa, Oklahoma 74101
1-800-752-9764

sales@pennwell.com
www.pennwell.com
www.pennwell-store.com

Printed in the United States of America
05 06 07 6 5 4 3

Library of Congress-Cataloging-in-Publication Data Pending

Johnson, David E.
Pile, Kathryne E.
Well Logging in Nontechnical Language, 2nd Edition

Managing Editor: Marla M. Patterson
Designed by Clark Bell & Robin Remaley

To my daughter, Kathleen

– D.E.J.

Contents

ACKNOWLEDGEMENTS
2ND EDITION

We gratefully acknowledge the help of the major well-logging companies and, in particular, Andy Shaw, Baker Atlas; Dave Gerrie and Furman Kelley, Halliburton; Lisa Silipigno and Dwight Peters, Schlumberger; Dale Walker, Tyco Production; and Edward Schaul, Schlumberger, retired. Without their help and generous support, this book would not have been possible.

A Word to the Reader

If you are interested in well logging, or if you work with well logs but don't know much about them, then this book is for you. *Well Logging in Nontechnical Language, 2nd Edition,* is an elementary yet practical text written for secondary users of logs—bankers, landmen, geology and engineering technicians, clerks, assistants, and secretaries—who come across logs in their daily routines and would like to make sense of those wiggly lines.

Before we begin, though, let's assess your needs. What is your interest in logging? Why do you want to know about logs? How much do you want to know? What are you going to do with the information? Most important, what can you expect to learn from this book? Will it make you an expert in log interpretation, or will it just make you dangerous?

Obviously, only you can answer the first couple of questions. If you have a passing or casual interest in logs, you've come to the right place. Maybe you've found yourself seated across a conference table from someone waving a log in one hand and a cigar in the other, swearing that his log is proof of the opportunity of a lifetime. On the other hand, maybe you've filed and handled logs day after day with only a vague idea about what they reveal. You may not want to become a petroleum geologist, engineer, or log analyst; however, you do want a solid background in logging so you can make more competent business decisions. Again, this book will help you.

What will you learn from this book? Well, after reading it, you'll be conversant with the main types of logs in use today—mud logs, open-hole and cased-hole wireline logs, computer-generated logs, and measurement-while-drilling logs. You'll be familiar enough with the more common types of open-hole logs to recognize productive zones and wet zones in simple cases. In addition, you'll know where to go for help with the more difficult cases. Though you won't be an expert, you'll know enough to ask the right questions. You'll also know when you have enough information to make a decision and, more importantly, when you don't.

This book is not the last word on the subject. It is a rudimentary, introductory explanation of a highly complex and technical subject. Use it as such. If you need to make an important decision based on the information contained in a set of logs, get help. Whenever you have money riding on the correct interpretation of a set of logs, seek expert advice from a consulting log analyst or from a log analyst/salesperson working for one of the major logging companies.

A second word of caution concerns the use of the logs. We can seldom measure directly the substance we are looking for (oil or gas). Instead, we make inferences and best guesses based on sophisticated measurements of other parameters. From these inferences, we formulate our interpretations. But think about it for a minute. If logs always succeeded in their predictions, if people interpreted them perfectly, and if logging tools never malfunctioned, logging companies wouldn't need the escape clause that is attached to all of their interpretations.

What is the escape clause? In essence, it states that we live in an imperfect world, full of well-intentioned but sometimes inept people; that machines and electrical instruments sometimes fail; and that occasionally interpretations will be wrong. Companies tell you this to underscore the high risk in any drilling venture. Logging companies and log analysts (and authors) are entirely blameless for any losses incurred as a result of these failures. We agree with these sentiments.

A book such as this one, which tries to translate a very technical subject into layman's terms, can never be as precise or exact as a technical treatment of the same subject matter. This book is a compromise between rigorous exactness and oversimplification. We hope we have hit a middle ground where the explanations are correct but simplified. We have often omitted or greatly abbreviated tool design principles and acquisition procedures and have presented only a couple of the many methods of interpretation available today. As we stated previously, we are not trying to make log analysts out of our readers; rather, we want to give them an appreciation for the process.

This book is aimed primarily at the petroleum industry. Oil and gas well logging accounts for most of the logs that are run. Nonetheless, there are several other branches to the logging family tree. A growing segment consists of using logging to evaluate mineral deposits for mineral exploration. Logs also play a role in geotechnical engineering, such as the study of the famous San Andreas fault; in environmental impact evaluation and monitoring of waste disposal wells; and in scientific investigation (many logs have been run for the federal government in monitoring and evaluating shot holes for underground atomic weapons testing). The logs used in this type of work are generally the same as those used in petroleum logging; so although your particular application may not be mentioned, this book describes the logs that you might use.

The examples in this book are mainly from the United States—not because they are unique, but because they were handy. An example is just that—an example, illustrating a point. It is not necessary to cover every geologic province in order to apply the methods described here to your particular part of the world. The units of measurement used in the book are those that were used for the log examples: English units. Many logs are run in metric units, but this should be of little importance to the reader because units of measurement are always noted on the log heading and scales.

Regardless of your interest in logs, the part of the world where you want to use them, or the measurement system that you prefer, this book should get you started in understanding petroleum well logs.

David E. Johnson
Kathryne E. Pile

1

Introduction to Logging

Just what are well logs and how did they get their name? One story goes something like this. When the oil industry was getting started around the close of the 1800s, many sailors were out of work. (Curiously, the sailors were unemployed because the fledgling oil industry and kerosene were eliminating the need for whale oil.) Since the sailors were used to working at heights and with rigging, they were naturals at scaling the tall oilfield derricks.

Along with the influx of sailors came many of their nautical expressions. That's why the drilling derrick and its equipment are called a *rig*, the derrick is a *mast*, the changing room is called a *doghouse*, and the records are kept in the *knowledge box*. The term *log* is another of these merchant marine expressions.

Nearly everyone has heard of the ship's log kept by the captain. It's a chronological record of what happens aboard a ship. The record of what occurs on a drilling rig is the driller's log. Since oil companies are interested in what happens as a bit drills deeper into the earth, the driller's log is usually recorded by depth rather than by time.

In the early days of the industry, the driller's log was practically the only information available about the subsurface formations. On the driller's log were recorded the types of rock brought up from the borehole, how many feet per hour the bit was drilling, oil or gas flows, equipment breakdowns, accidents such as stuck drillpipe, and any other occurrence that might have a bearing on evaluating the well. Today, *log* has stretched to mean any data recorded versus depth (or time) in graph form or with accompanying written notes.

When someone mentions a log, he or she is usually referring to records run on an uncased wellbore using an electric wireline logging truck and tools (Fig. 1–1). Logs can also refer to the driller's log, mud logs, computer-generated logs, and MWD (measurement while drilling) logs.

Fig. 1–1. *Electric logging truck that carries an onboard computer and enough cable to log wells as deep as 25,000 ft. (courtesy Schlumberger).*

WHY DO WE RUN LOGS?

What are we trying to accomplish with a log? What does it tell us that is so important?

One of the advertising slogans of Schlumberger (pronounced slumber-jay) during the 1960s was "...the eyes of the oil industry." This slogan aptly describes the importance of logging. Geologists and engineers literally work blindly when they try to imagine what is happening at the bottom of a well. Layers upon layers of sediments have amassed over the years and have been deformed and altered so much that we can't guess exactly what lies beneath our feet.

Before logging, drillers had only the information from their driller's logs and the behavior of nearby, or *offset,* wells. This information was and is important and useful, but it still left a lot to "by guess and by golly." Electric wireline logs have turned on the light for the petroleum geologist and engineer. In particular, they provide information in areas such as these:

- depth of formation tops
- thickness of formations
- porosity
- temperature
- types of formations encountered (shale, sandstone, limestone, dolomite)
- presence of oil or gas
- estimate of permeability
- reservoir pressures
- formation dip (the angle the formation makes to the horizontal and its direction)
- mineral identification
- bonding of cement to the casing
- amount and kind of flow from different intervals in a producing well

The list goes on and on, and new logs as well as new uses for old logs are being developed continually.

But the real reason for running logs is to determine whether a well is good or bad. A good well is commercially productive—it produces enough oil or gas to pay back its investors for the cost of drilling and leaves a profit. A bad well is not commercially productive. Logs help us make this determination.

By the time a log can be run, thousands of dollars have been spent for leases, possibly for seismic studies, and for drilling. However, thousands of dollars more are still to be spent to complete the well—running the casing, cementing, perforating, testing, setting production tubing and packers, and installing wellhead equipment and surface production facilities. If a company can determine that a well won't be productive before it spends thousands of dollars on the completion costs, it will minimize its loss. As in poker, there's no sense in throwing good money after bad.

Logs help us determine whether the formation we are penetrating contains commercial reserves of oil or gas, thus minimizing costs on bad wells. On good wells, logs also show us where the oil or gas may lie, how much there is (reserves), and whether more than one zone is productive.

WHO USES LOGS AND WHY?

Practically everyone in the oil industry uses logs at one time or another (Fig. 1–2). And logs are certainly used by everyone involved in the decision-making processes necessary in drilling and completing a well.

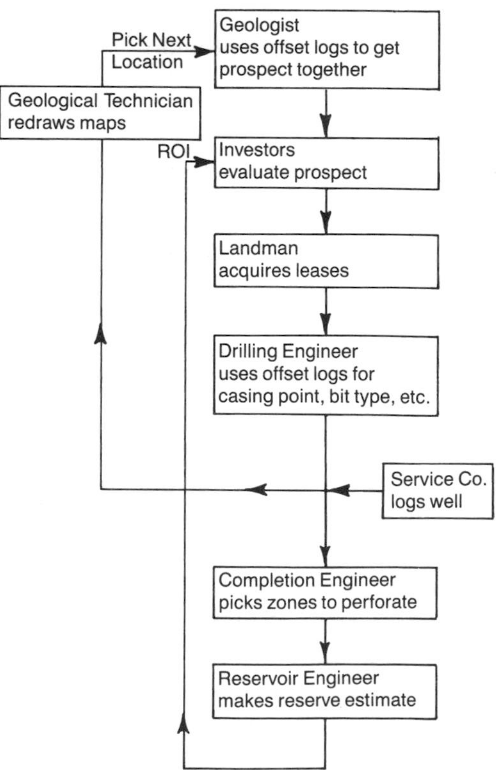

Fig. 1–2. *Logs are used by practically everyone in the oil industry.*

Logs are used in nearly every phase of the exploration and production process. Let's take a look at a drilling deal put together by an independent (any oil-related company not connected with a major, meaning a mega-oil company with production, refining, and marketing capabilities). An independent can vary in size from multimillion-dollar companies with hundreds of employees to a group of several people with expertise in different areas, or even just one person. An independent may also be a coalition of individuals formed for just one drilling deal. Included somewhere in the group are geologists, landmen, moneymen, and engineers.

First, the *geologist* evaluates an area. The evaluation is based on seismic data, existing logs, nearby well data, imagination, and intuition. Armed with this information, the geologist draws structural maps of the area and recommends how to develop the prospect, or *play*.

The *landman* is primarily responsible for obtaining the leases necessary for the geologist's play. The landman doesn't need to know how to read the logs as accurately as a log analyst. However, he or she must have a working knowledge of logs to discuss prospects with landowners, bankers, and geologists.

The data gathered by the geologist (including the logs) might be used next by the *moneymen* who sell the deal to a banker or investors. The parties that buy into the deal have a working interest, i.e., they invest money in the hope they will reap profits. To protect the investment, bankers or investors often evaluate the log data independently, using in-house or consulting log analysts. After the well has been drilled and the logs have been run, opinions within the group may differ on whether to spend more money on a completion attempt or to plug and abandon the well. Whether it's yours or your company's money at risk, it's good to know enough about log interpretation to decide which way to go.

The *drilling engineer* drills the well on the basis of log information from nearby wells. From this information, the engineer decides the kind and weight of mud to use, the types of formations to be encountered, the kind of drill bits to use, where to set casing, and how long it will take to drill the well.

The *completion engineer* relies heavily on logs to determine which zones are probably productive and exactly where the casing should be perforated. On the basis of information from daily reports, the mud log, and various open-hole and cased-hole logs, the completion engineer will perforate, test, treat, and finally put the well on production.

The *reservoir engineer* uses the open-hole logs to make the initial calculation of reserves (the amount of producible oil or gas). These reserve calculations are updated periodically from production data, pressure buildup tests, and possibly other logs run later in the life of the well.

Included in this string of people who use logs are the *geological technician* who drafts the structural maps, the *royalty owner* who wants to know why one well wasn't as good as a neighbor's, the *mud salesperson* making a pitch for a better mud system on the next well, and the *accountant* who calculates the net worth of the company's assets indirectly using well-log data to make an evaluation. Many people depend on the interpretation of well logs. That's why it's good to be as knowledgeable as possible about logs.

The first step toward learning about logs is knowing their components and how to read them, so let's turn to Chapter 2 and get started.

2

Reading Logs

In the next chapters, we'll look at several types of logs used in the petroleum industry. Some measure the resistivity of formations, others determine porosity, and still others determine types of minerals present. But before we study what these various logs tell us about the earth's formations and the presence of oil or gas, we need to know where to locate and how to read the five major sections of a log: the header, the main log section, the inserts, the repeat section, and the calibrations.

THE HEADER

When you are handed a log, the first thing you usually see is a short section of text at the top of the log. This section is called the *heading* or *header* because, as the name implies, it is attached to the top, or head, of the log. The header contains useful and often critical information. As you read the list below, note the corresponding location on Figure 2–1:

1. Logging company
2. Operating company (operator)
3. Specific well information
   – well name or number
   – lease or field name

- legal location (where the well is located geographically, often a certain section, township, and range)
- elevation of ground surface above sea level and usually elevation of the rig floor or the kelly bushing
- date when the logs were run
- total depth of the well at the time of logging
- miscellaneous information (drilling mud properties, bit size, casing size, depth)

4. Type or kind of log run
5. Other logs or surveys run on the well
6. Equipment information
   - tool serial numbers
   - tool spacings
   - truck number
   - office that supplied the truck
7. Personnel information
   - person who recorded the log
   - person who witnessed the log (company man)
8. Remarks section for noting any unusual conditions or occurrences during the logging job
9. Log scales and curve identification

A first rule to follow when interpreting any log is to examine the heading carefully. Why? The most obvious reason is to be certain you're reading the logs from the appropriate well. Then check the technical data contained in the heading information to determine the type of log and other information about the well, such as mud data (resistivity, water loss, weight, and viscosity), bit size, depth of any casing, and total well depth. Look in the Remarks section to determine if anything unusual occurred during the logging run. Finally, see which logging company ran the log. Not all logging companies are equal in the quality of their work, and some of your decisions may be influenced by how much confidence you have in the log readings. All of this information is important to your interpretation of the logs and can help you decide what to do with the well.

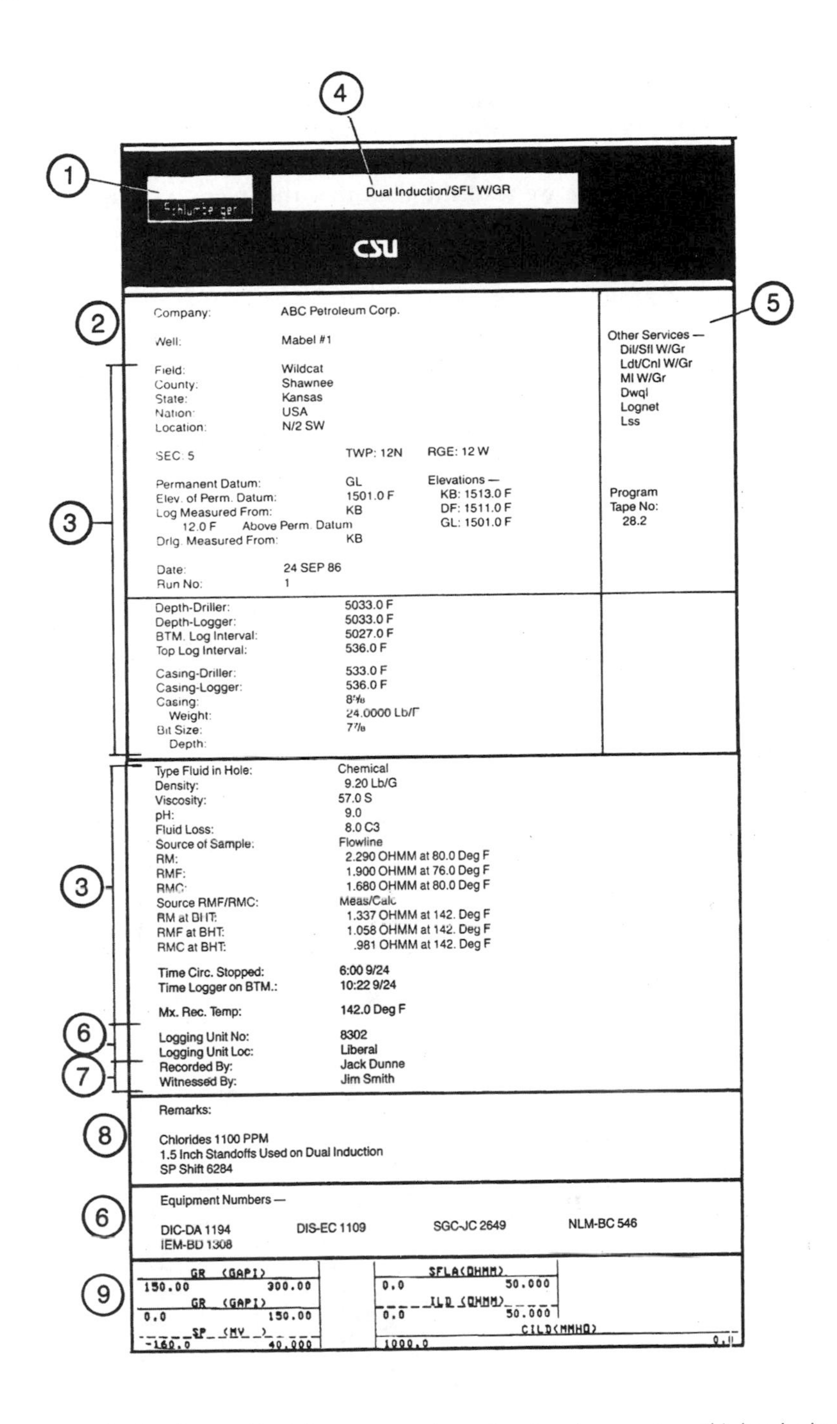
Schlumberger

Dual Induction/SFL W/GR

CSU

Company: ABC Petroleum Corp.
Well: Mabel #1
Field: Wildcat
County: Shawnee
State: Kansas
Nation: USA
Location: N/2 SW

SEC: 5 TWP: 12N RGE: 12 W

Permanent Datum: GL
Elev. of Perm. Datum: 1501.0 F
Log Measured From: KB
12.0 F Above Perm. Datum
Drlg. Measured From: KB

Elevations —
KB: 1513.0 F
DF: 1511.0 F
GL: 1501.0 F

Date: 24 SEP 86
Run No: 1

Other Services —
Dil/Sfl W/Gr
Ldt/Cnl W/Gr
Ml W/Gr
Dwql
Lognet
Lss

Program Tape No: 28.2

Depth-Driller: 5033.0 F
Depth-Logger: 5033.0 F
BTM. Log Interval: 5027.0 F
Top Log Interval: 536.0 F

Casing-Driller: 533.0 F
Casing-Logger: 536.0 F
Casing: 8⅝
Weight: 24.0000 Lb/F
Bit Size: 7⅞
Depth:

Type Fluid in Hole: Chemical
Density: 9.20 Lb/G
Viscosity: 57.0 S
pH: 9.0
Fluid Loss: 8.0 C3
Source of Sample: Flowline
RM: 2.290 OHMM at 80.0 Deg F
RMF: 1.900 OHMM at 76.0 Deg F
RMC: 1.680 OHMM at 80.0 Deg F
Source RMF/RMC: Meas/Calc
RM at BHT: 1.337 OHMM at 142. Deg F
RMF at BHT: 1.058 OHMM at 142. Deg F
RMC at BHT: .981 OHMM at 142. Deg F

Time Circ. Stopped: 6:00 9/24
Time Logger on BTM.: 10:22 9/24

Mx. Rec. Temp: 142.0 Deg F

Logging Unit No: 8302
Logging Unit Loc: Liberal
Recorded By: Jack Dunne
Witnessed By: Jim Smith

Remarks:

Chlorides 1100 PPM
1.5 Inch Standoffs Used on Dual Induction
SP Shift 6284

Equipment Numbers —

DIC-DA 1194 DIS-EC 1109 SGC-JC 2649 NLM-BC 546
IEM-BD 1308

GR (GAPI) 150.00 300.00
GR (GAPI) 0.0 150.00
SP (MV) -160.0 40.000

SFLA(OHMM) 0.0 50.000
ILD (OHMM) 0.0 50.000
CILD(MMHO) 1000.0 0.0

Fig. 2–1. *The header provides information on type of well and parameters. This header has been cut in half and the top half flipped for illustration purposes.*

MAIN LOG SECTION

Just below the header is the main body of the log, which looks like a very long graph. Here we read the data that the logging equipment transmits to the surface. In this section we will learn to read both vertical and horizontal scales.

Vertical Scale

The vertical or long axis of the log represents the depth of the well and records the exact depth at which formations occur (Fig. 2–2). The *depth track* or *depth column* is the vertical space with numbers near the center of the log. Depth numbers are printed in this space in multiples of 100 ft. of depth and correspond to the horizontal depth lines on the graph.

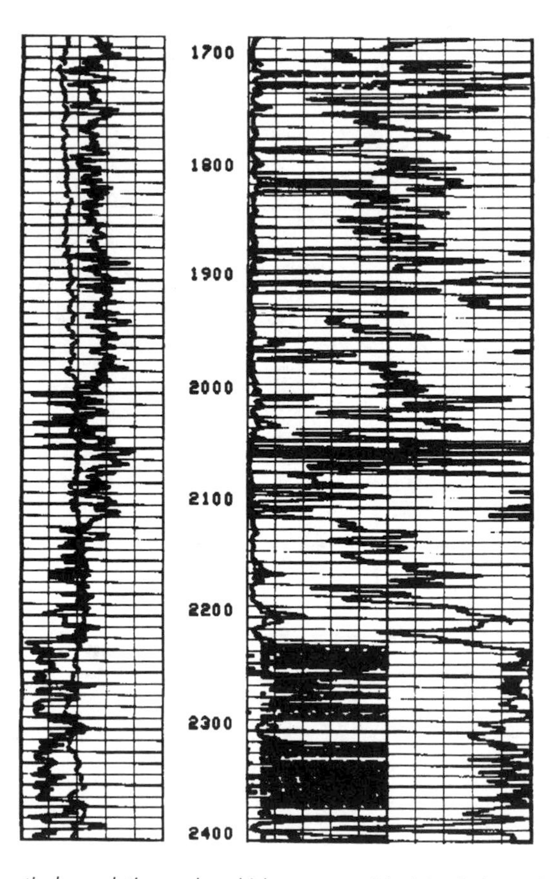

Fig. 2–2. *This vertical correlation scale, which measures 1 in./100 ft. is used to compare formation depths of nearby wells.*

Depth scales are always linear—that is, the division marks are of uniform spacing, just like the division marks on a ruler. The depth scale on a log is usually either 1, 2, or 5 in. per 100 ft. of hole. This means if you laid a ruler on the depth scale of an actual log and measured between any two consecutive hundred-foot depth numbers (such as 1600 to 1700 ft.), the distance would be 2 in. for a 2 in./100 ft. scale and 5 in. for a 5 in./100 ft. scale. These scales are usually referred to as the 2-in. log or the 5-in. log, respectively.

In addition to the dark horizontal lines at the 100-ft. depth numbers, the 1-in. and 2-in. scales have both 50-ft. and 10-ft. depth lines. Look at Figure 2–3 and find the depth 1650 ft. (A). Now find the depths 1680 ft. (B) and 1738 ft. (C). You had to mentally divide, or interpolate, the distance between the 1730-ft. and the 1740-ft. depth lines and estimate where 1738 ft. is.

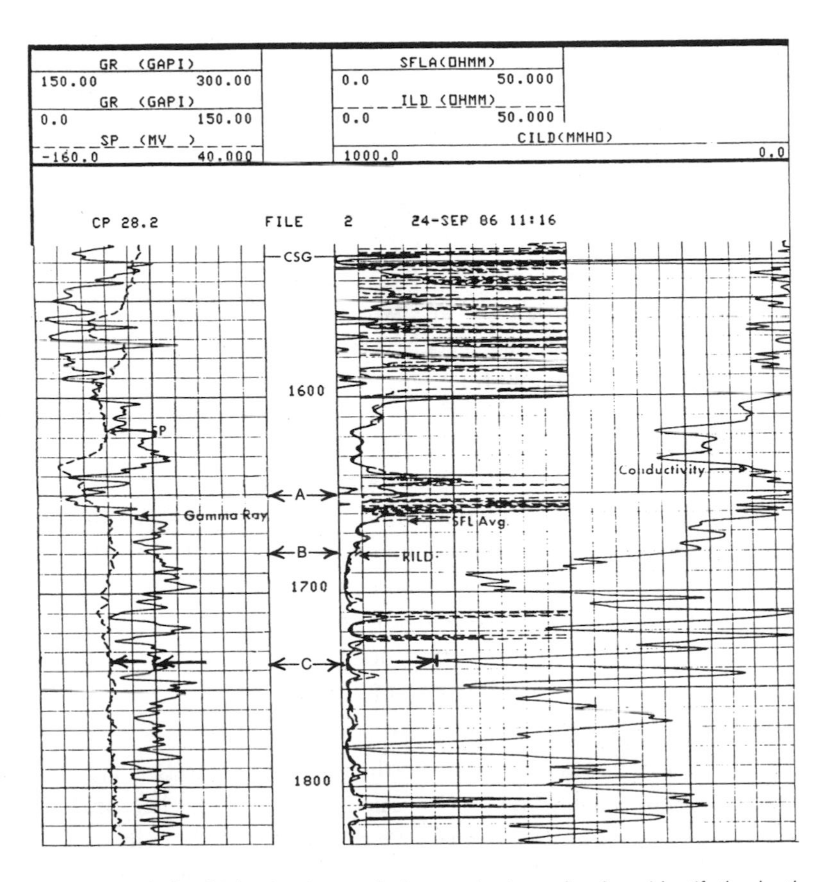

Fig. 2–3 *The scale for this log is 2 in./100 ft. For practice in reading logs, identify the depths at points A, B, and C.*

The depth lines on the 1 in./100 ft. scale are divided the same as on the 2-in. scale: 10-ft., 50-ft., and 100-ft.. lines and 100-ft. depth numbers. If we had a 5-in. log, we would have 100-ft. lines at the depth numbers, 50-ft. lines of the same weight (degree of darkness or width) as the 100-ft. lines, 10-ft. lines somewhat thinner than the 50- and 100-ft. lines, and 2-ft. lines the thinnest of all. By having 2-ft. lines and the expanded scale of the 5-in. log, we could easily read depths to 6 in.

The 2 in./100 ft. and 1 in./100 ft. scales are called correlation scales. Geologists use the correlation scale to compare logs between several wells over large intervals of formation. The 2-in. scale is usually used to correlate with one or two nearby offset wells, while the 1-in. scale is often used to construct cross-sections over several miles of surface and many thousands of feet of formation. The 5 in./100 ft. scale is called the detail scale because more features can be noted with this expanded scale than with the correlation scales.

In addition to these three common depth scales, other special scales are occasionally seen. *Super detail scales,* 10 in./100 ft. or 25 in./100 ft., are used most often with micrologs or fracture identification logs and are run over short intervals of the hole.

To the left of the depth track is track 1 (see Fig. 2–3). This track is often called the SP *(spontaneous potential) track* or the *GR (gamma-ray) track* after the two curves, or measurements, that are most commonly recorded there. To the right of the depth track are two more measurement tracks, track 2 and track 3. Various parameters such as resistivity, porosity, and sonic velocity are recorded in these two tracks, as discussed in later chapters.

Horizontal Scale

We have already talked about the vertical or depth scale. Now let's consider the horizontal scale, which is the measurement scale. To put a value or number on the various formation parameters we are measuring, such as resistivity and porosity, we must assign a measurement scale to the appropriate measurement track. This horizontal scale may take one of several forms. So let's talk about scales and graphs in general and then come back to horizontal log scales and try our hand at reading some.

Figure 2–4 is a very simple log with measurement curves in tracks 1 and 2. Note that the curves are labeled at the bottom. The GR curve in track 1 is represented by a solid line that is scaled from 0 on the left, at division 0, to 100 on the right, at division 10. We can determine how many GR units are represented by each division by dividing 100 (the number of units) by 10 (the number of divisions). Each division is worth 10 GR units.

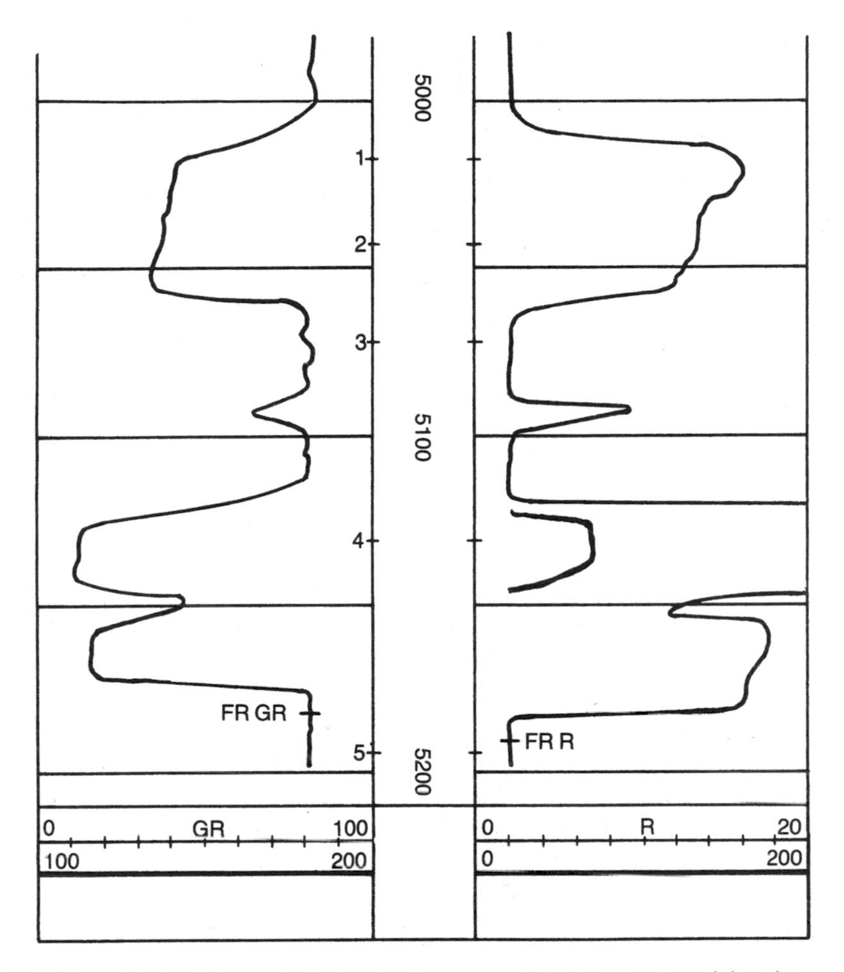

Fig. 2–4. *In this simple log, note the two tracks—one scaled into 100 units and the other into 20 units.*

Now look at the R *(resistivity)* curve in track 2. This curve is also solid, but it could be short-dashed, long-dashed, or dotted, and the curve weight could be light or heavy. It is scaled from 0 at division 0 of track 2 to 20 at division 10 of track 2. The R curve has a backup curve, shown by a heavy solid line, scaled from 0 to 200. This curve, a one-tenth backup curve, does not print or show on the log until the primary curve (scaled 0–20) goes off the scale. Note that 10 divisions of track 2 are the same as 0 divisions of track 3. Often, curves will be scaled completely across both tracks 2 and 3; in other situations, such as this one, the curve cuts off (disappears) at division 10.

Exercise 1

Let's get some practice in identifying scales and reading values from the measurement curves. This is called "reading the log." In Figure 2–4, how many R units are there per division? If you said 2 for the primary curve and 20 for the backup curve, you're correct. Now let's practice reading these curves. First, make a table like Table 2–1 on a piece of paper.

Point	Depth	GR	R
1	5020	40	16
2			
3			
4			
5			

Table 2–1. *Exercise 1 for reading measurement curves. Use along with Figure 2–4.*

Read the depth of each point and the curve values at that point. For the first point we get 5020 ft. The GR is four divisions from 0, so it must have a value of 4 times 10, or 40. Write 40 under GR next to the depth 5020. For the R reading, you should have 8 divisions times 2 R units per division, or 16. Go ahead and fill in your table with the other readings. The answers are at the end of this chapter.

Be sure to notice one other point on Figure 2–4. At the very bottom of the curves is a short line labeled FR GR and FR R. *FR* stands for the first reading of that particular curve. **Never** read a curve below this point. Even though the curve might appear to be recording a measurement, it's not; the logging tool is resting on the bottom of the well. The reason why the FR R is deeper than the FR GR is that the gamma-ray tool was above the resistivity tool on the wireline and could not be lowered as far.

Now look at Figure 2–5 for some more practice in reading log values. Once again, start by looking at the curve identification at the bottom of the log. Note that track 1 has a GR curve scaled from 0–150 with a backup scale of 150–300. This means each division will be worth 15 units on the GR scale.

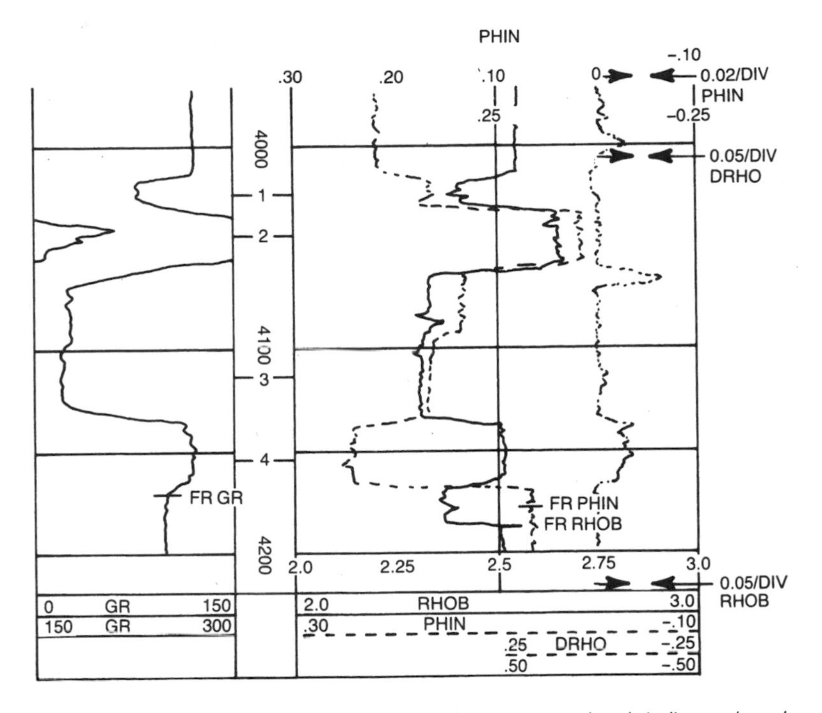

Fig. 2–5. *In this simple three-track curve, the space between arrowheads indicates the value of one division for each curve. Try your skill by reading the log at points 1–4.*

Tracks 2 and 3 are much more complicated than in the last example. Here we have three different curves with different scales and curve codings. The solid curve is labeled RHOB (pronounced "row B") and is scaled from 2.0 at division 0 of track 2 to 3.0 at division 10 of track 3. (Ten divisions of track 3 are the same as 20 divisions of track 2.) To find out how many units of RHOB each division is worth, subtract the reading at 20 divisions (3.0) from the reading or scale value at 0 divisions (2.0) and divide by 20. The answer is 0.05 units per division (3.0 – 2.0 = 1.0; 1.0/20 divisions = 0.05).

Where is the 0 porosity line for the PHIN (pronounced "fee N") curve? To determine this, you have to find out how many PHIN units there are per division. Note that the scale reads backwards—that is, the higher values are on the left and the lower values are on the right. In fact, division 10 of track 3 (this is the same as division 20, track 2) is labeled –0.10, a negative value. How do we determine the units per division for this curve? First, we subtract the value for PHIN at division 0 (0.30) from the value at the right edge of track 3 (or division 20, track 2). This value is –0.10. Therefore, the PHIN units per division are –0.10 – 0.30 = –0.40. Divide –0.40 PHIN by 20 divisions: –0.40/20 = –0.02 PHIN/division. The minus sign means the scale increases from right to left instead of from left to right. So where is 0 PHIN? It's at division 5, track 3 (division 15, track 2). The RHOB and DRHO (called "delta rho") scales are both 0.05 per division.

Exercise 2

Now let's get some more log-reading practice by recording the values of GR, PHIN, RHOB, and DRHO at points 1, 2, 3, and 4. Set up a table as you did for the previous exercise.

Point	Depth	GR	PHIN	RHOB	DHRO
1	4024	75	0.17	2.40	0.0
2					
3					
4					
5					

Table 2–2. *Exercise 2 for reading measurement curves. Use along with Figure 2–5.*

The horizontal scale of track 1 is always linear, that is, the increments are of uniform size like a ruler's. However, tracks 2 and 3 are often printed in a variety of formats. Besides being linear, both tracks may be logarithmic, or the two tracks may even be a split scale in which track 2 is logarithmic and track 3 is linear.

Logarithms are based on powers of 10. Each cycle of numbers is 10 times as large as the preceding cycle. Think of logarithms in terms of money (see Fig. 2–6). If the lowest value shown is 1¢ (or 1 penny), the first cycle is scaled from 1¢ to 10¢, which is the same as 1 dime. The second cycle is scaled from 10¢, or 1 dime, to 100¢ (1 dollar bill or 10 dimes.) The third cycle is scaled from 100¢ (1 dollar bill) to 1000¢ (one 10-dollar bill or ten 1-dollar bills). For the computer literate among us, a logarithmic scale is like file compression. The scale allows a large range of data to be displayed in a small space.

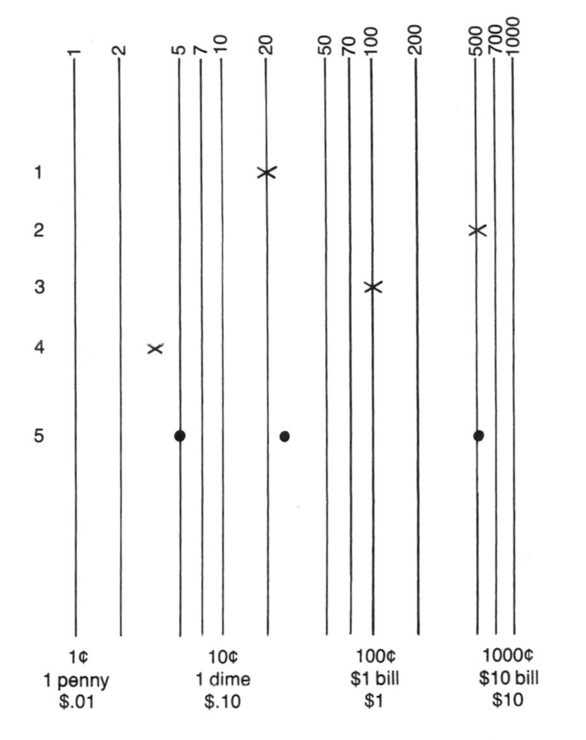

Fig. 2–6. *Read the graph using the guidelines in the text to learn how to read logarithmic scales.*

Exercise 3

To understand logarithms better, set up another table and record the values at the four points indicated on Figure 2–6.

Point	Value	Monetary Value
1	20	20¢
2		
3		
4		
5		

Table 2–3. *Exercise 3 for reading logarithmic measurement curves. Use along with Figure 2–6.*

On Figure 2–6 read the values at points 1, 2, 3, and 4. The value at point 1 is 20¢. (See the scales at the bottom of the figure.) Now read the points for 2, 3, and 4. Notice that point 5 is a little different; it shows how a log scale may be used to give a more precise reading of a value. If we try to read just the point at the far right, we estimate something between 500 and 550; but by using the other two points, we can refine our reading to get 525 (500 + 20 + 5).

Exercise 4

To get some practice with a real log, note Figure 2–7. Here we have dual induction-SFL curves recorded on a logarithmic scale in tracks 2 and 3. Note that the scale starts at 0.20 instead of 0.10, but otherwise it is just the same as the example in Figure 2–6. Read the three curves at points 1 and 2; check your answers at the end of the chapter.

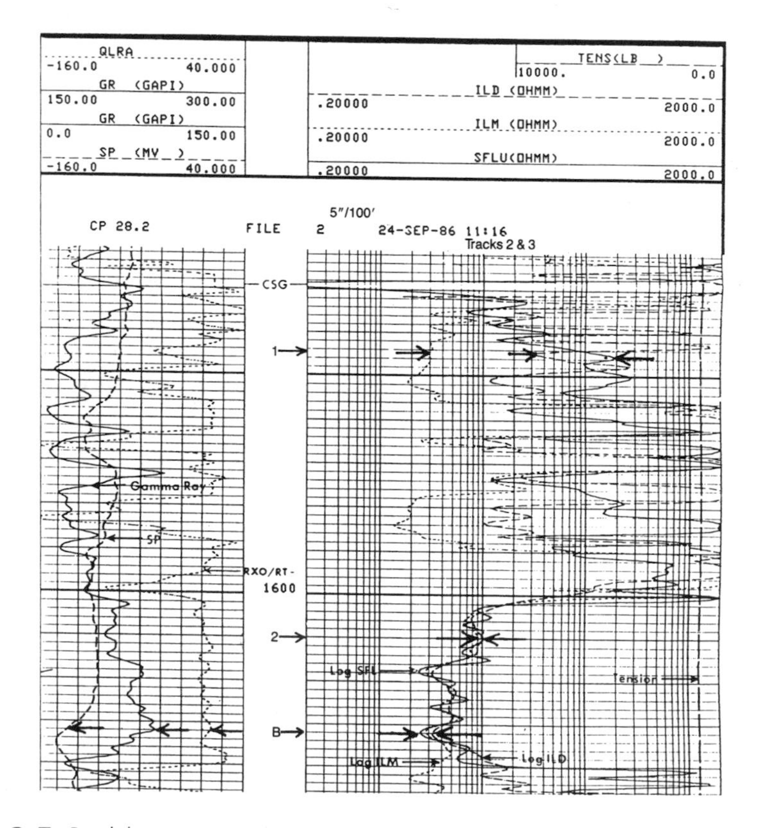

Fig. 2–7. *Read the curves at points 1 and 2 on this logarithmic scale and check your answers at the end of the chapter.*

Other common presentations record the depth track at the left edge of the log and divide the space to the right into four tracks. This presentation is often used with computed or computer-interpreted logs. A computer-interpreted log has been generated from combinations of two or more logs.

INSERTS

Between each major section of the log and at the end of the log are inserts. Inserts label the measurement scales and identify each curve (measurement). Curves (also called *traces, readings,* or *measurements*) may be printed as solid, long-dashed, short-dashed, or dotted lines. All of these lines may be either heavy or light.

For an example, let's look at Figure 2–8. Observe the upper part of the insert that labels the curves at the bottom of the 2-in. log scale. (**Note**: Don't be distracted by the alphabet soup of names and labels that are probably meaningless to you right now. The terms are defined later on. For now, just notice that each row has a name and a special coding to help identify it on the log). The GR curve is labeled in track 1 with a scale of 0–150 units and is a heavy solid line. The SP curve, also in track 1, is labeled as a dashed line with a scale of –160 to +40 (20 units per division). In track 2, the curve labeled SFLA is shown as a solid line with a scale of 0–50 ohms. The ILD curve is shown as a dashed line in track 2 with a scale of 0–50 ohms. The CILD curve crosses both tracks 2 and 3. It has a scale of 1000–0 millimhos and is shown as a solid line.

The insert identifies all of the curves on each section of the log so the various curves can still be recognized as the presentation changes from the 2 in. to the 5 in./100 ft. scale. At the top of the 5-in. log, all of the curves are identified again. Here we have some additional curves, and tracks 2 and 3 are now in the logarithmic format. See if you can properly label and locate all of the curves shown for the 5-in. scale.

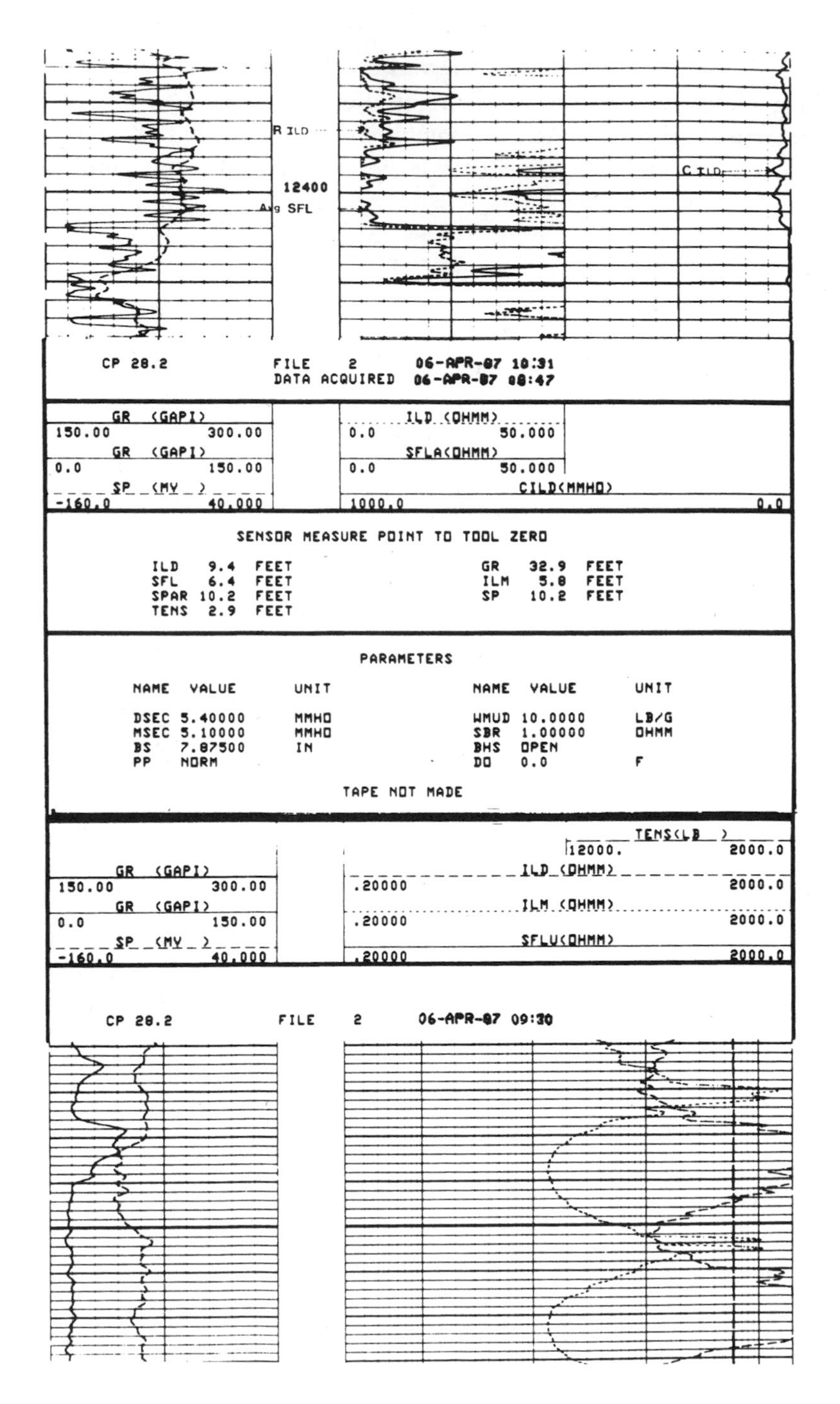

Fig. 2–8. *Inserts list scales and identify curves.*

REPEAT SECTION

The repeat section verifies that the measurements on the main log repeat themselves, meaning they read the same at least twice. Therefore, the logging equipment was working properly when the log was made (Fig. 2–9). Always lay the repeat section alongside the main log section and check at several different places to ensure the readings repeat. If they do not, rerun the logs with different equipment. If this is impossible, use the log information with caution. Radioactive measurements always show a certain amount of variation because of the random nature of radioactive decay. Other measurements, such as resistivity and sonic data, should repeat very closely. The repeat section is attached below the main body of the log.

CALIBRATIONS

The final section of the log shows the before-and-after survey (logging) calibrations. These calibrations verify the tools were adjusted properly before the log was run and they were still in adjustment at the end of the logging job. For our purposes, we will assume the tools were all properly calibrated. If you ever suspect a log was not calibrated properly or the tools changed during logging and became uncalibrated (out of tolerance), then contact the logging company, question its procedures, and have the company verify the tools were properly calibrated before and after the survey.

COMPREHENSION CHECK

To see how well you understand this chapter and especially how well you read log values, turn back to Figures 2–3, 2–7, and 2–9 and read all of the curves and depths at points A, B, and C. Check your answers at the end of the chapter. If you don't do too well, go back and reread the sections you had trouble with.

We now know how to identify and read the major components of a log. Let's turn to Chapter 3 to see what these curves are telling us so we can put everything together and interpret our logs.

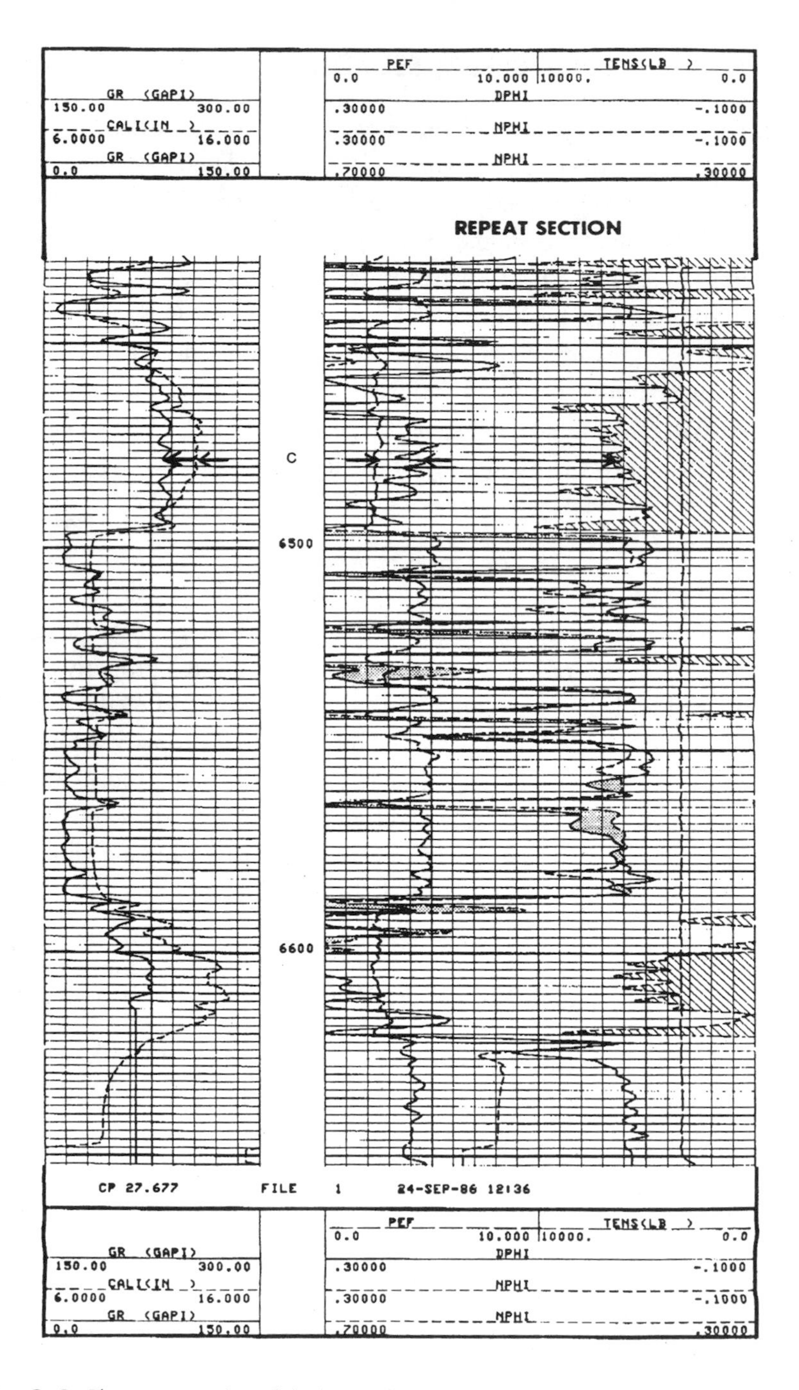

Fig. 2–9. *The repeat section of the log verifies that the logging equipment was working properly and that the measurements repeat themselves.*

Answers to Exercises

Exercise 1

Point	Depth	GR	R
1	5020	40	20
2	5043	36	13
3	5075	82	2
4	5131	11	70
5	5194	*	*

**These points are below the FR (first reading). Never read a curve here; it's not a valid measurement.*

Exercise 2

Point	Depth	GR	PHIN	RHOB	DRHO
1	4024	75	0.17	2.40	0.0
2	4045	180	0.015	2.65	0.01
3	4114	20	0.17	2.31	0.02
4	4155	120	0.25	2.52	0.06

Exercise 3

Point	Value	Monetary Value
1	20	20¢
2	500	500¢ or $5
3	100	100¢ or $1
4	3	3¢
5	525	525¢ or $5.25

Exercise 4

Point	Depth	ILD	ILM	SFLU
1	1545	32	3	200
2	1610	10	9	1

Answers to Comprehension Check

Fig. 2–3

Point	Depth	GR	SP	SFLA	ILD
A	1650	30	-120	150	—
B	1670	75	-92	3	5
C	1737	75	-100	2	2

Fig. 2–7

Point	Depth	GR	SP	ILD	ILM	SFLU
B	1632	85	-130	3.2	2.9	2.5

Fig. 2–9

Point	Depth	GR	CAL	PEF	DPHI	NPHI
C	6475	85	13.6	2.4	20.5	4

3

FORMATION PARAMETERS

Before we can begin interpreting logs, we first have to cover some basics. We need to know a little about the different kinds of reservoir rocks, how these reservoirs rocks are formed, the fluids contained in the rocks, and how the hydrocarbons are distributed inside the reservoir rock. We also need to examine direct versus indirect measurements and learn about one of the earliest and still very useful measurements that we make—resistivity.

TYPES OF SEDIMENTS

As you are probably aware, oil and gas deposits are contained within sedimentary rocks, one of the three major types of rock. These rocks, formed by sediments deposited in layers on the bottom of rivers, lakes, and oceans, can be classified into three categories: clastics, evaporites, and organics.

Clastics, from the Greek word *klastos,* meaning "broken," are formed from the fragments of other rocks. The most common example of this type of rock is sandstone. Individual sandstone grains are fragments of other rocks that were weathered, worn, crushed, and tumbled by wind and water until they were deposited somewhere long enough to be buried by other sediments. These sediments are then cemented by chemical action, heat, and pressure. Sandstones (or sands, as they are commonly called) are the most important reservoir rock; most hydrocarbon accumulations occur in sandstones.

Clastic sediments are differentiated by the size of the sediment grains, or *clasts*, that form the rock. Sandstones have medium-sized grains, ranging from the size of beach sand to very fine, barely visible particles. *Conglomerates*, another kind of clastic sediment, have large grains—from very small pebbles the size of grains of rice to rocks bigger than a man's fist. Conglomerates are always very poorly sorted, i.e., there are a large variety of grain sizes in the rocks. *Shale* is yet another clastic rock; its grains are microscopic. In addition, shales contain various types of clay minerals.

Evaporites are formed when a saline body of water evaporates. As the salinity increases, certain chemicals precipitate and fall to the bottom, forming a layer. The most common evaporites are gypsum, anhydrite, and halite (rock salt). Halite is especially interesting to oil explorers because many oilfields have been associated with salt domes, especially along the Gulf Coast.

Most organic sediment is classified as a *carbonate*, which includes limestone and dolomite. Carbonates occur in many forms: extremely fine-grained micrites from limestone mud, chalk from the excrement of golden-brown algae, and reefal limestones from coral. Organic sediments are usually the skeletons of small marine organisms that sink to the bottom and accumulate. Over time, these organisms form layers that are often hundreds of feet thick. The layers are eventually buried, compacted, and cemented together, making a carbonate bed an important reservoir rock.

All sedimentary rocks have some water in their pore spaces (the space between the individual grains). This water is referred to as *formation water* and contains varying amounts of dissolved salts. The source of the formation water may be the lake, river, or ocean in which the original sediments were deposited or, as in the case of sand dunes, the source may be rainwater or later groundwater that migrates in as the formations are buried.

In Figure 3–1, we see grains of various kinds of minerals forming a sandstone. These grains have been cemented together by heat, pressure, and a cementing material such as carbonate and/or silica. Since they were either laid down in a watery environment or were later submerged, the sandstone grains normally will be water wet, i.e., water clings to the individual grains. Nearly all formations have this immovable, or irreducible, water. No matter how much oil or gas a formation may contain, the irreducible water remains.

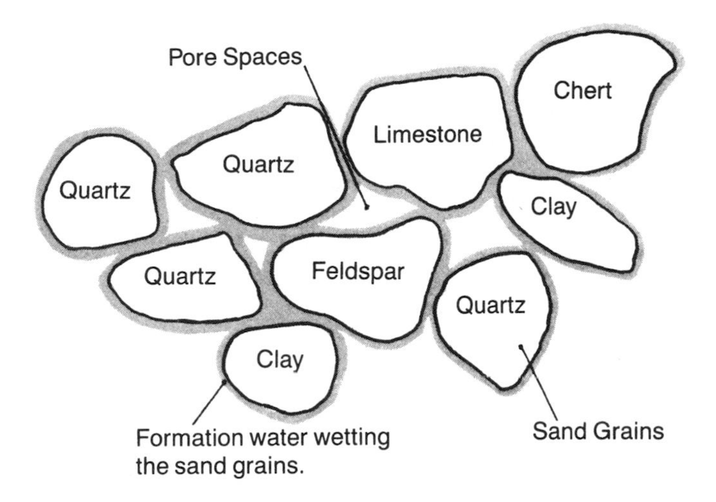

Fig. 3–1. *Schematic of sandstone grains. Various kinds of materials comprise the sand grains. Formation water (gray) coats the grains. The pore spaces left between the grains can hold other kinds of fluids, such as water, oil, or gas.*

POROSITY

When sediments are deposited and compacted, they do not form a solid mass of rock. Spaces exist between the grains (intergranular pores). The amount of space, or voids, as a percentage of the total volume of formation, is called the *porosity*, or nonrock volume. Formation fluids (oil, gas, or water) accumulate in the voids (Fig. 3–1). The larger the porosity, the more fluids a formation contains.Without porosity, a formation is of no interest to an oil explorer because there is no place in the rock for hydrocarbons to accumulate.

Intergranular Porosity

To gain a clearer understanding of porosity, let's perform an experiment. Take a box that measures 1 ft. on each side (volume = 1 ft.^{3}) and fill it with uniformly sized marbles (Fig. 3–2a). The space between the marbles is the pore space. We can determine how large the pore space is by using a measuring cup to pour water into the box. If we measure very carefully, we find we can add 0.476 ft.^{3} of water. This means the box of marbles has a pore space, or porosity, of 0.476 ft.^{3}, or 47.6%.

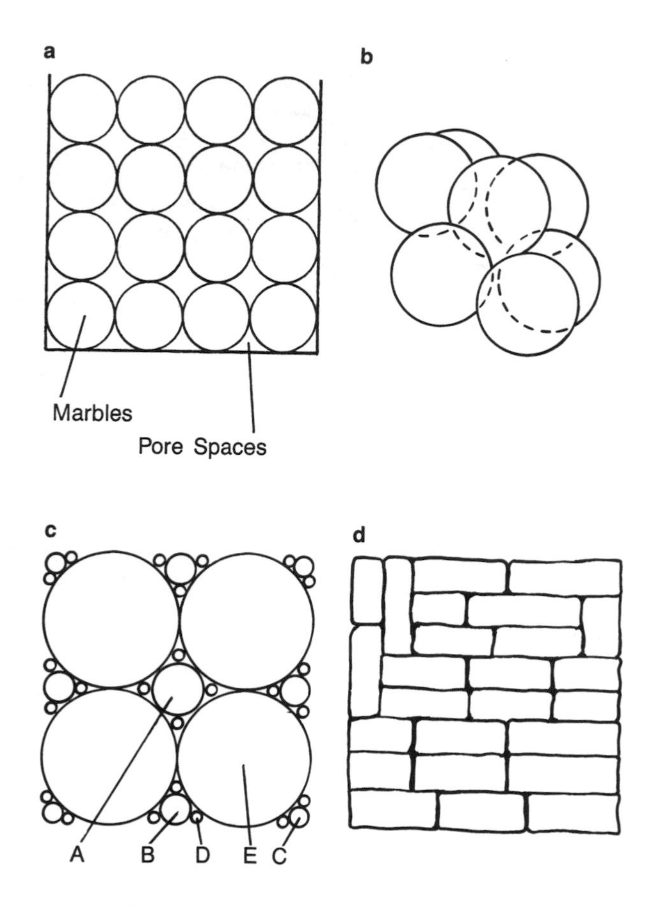

Fig. 3–2. *How porosity changes relative to grain shape and arrangement. (a) Maximum possible porosity; porosity ≈ 47%. (b) Rhombohedral stacking; porosity ≈ 26%. (c) Various sizes of grains (A–E) in a closely packed arrangement; porosity ≈ 10%. (d) Loosely stacked bricks; porosity ≈ 1%.*

The marbles in Figure 3–2a represent sand grains. Obviously, sand grains are not symmetrical like marbles, so this experiment represents the best, or limit, conditions. Look at the way the marbles are arranged in the box, stacked on top of each other. Sand grains seldom align themselves like this in nature because the arrangement is unstable. Here we have the largest possible porosity for a sandstone; so if we see a porosity reading greater than 47.6% on a log, we know we are not looking at a sandstone formation.

An odd but interesting mathematical fact is that the porosity will not change with the size of the marbles. Whether we put a 1-ft. wide marble in the box or carefully stack it full of 1-mm marbles, the porosity will be the same as long as the arrangement is the same. If you're good at simple geometry, you can prove this to yourself. The equation for the volume of a sphere is

$$V = 1/6\ \pi\ d^3,$$

where:

V = volume
d = diameter of the sphere
π = 3.1416

What happens if we alter the arrangement of the marbles? In Figure 3–2b the marbles are arranged in a rhombohedron—that is, they are stacked together the way you'd stack cannon balls. The maximum possible porosity now is 26%. This arrangement is stable and is more likely to exist in nature. We can say that a well-compacted, well-sorted (sand grains are essentially the same size) sandstone should have a maximum porosity of around 26%.

If we vary the size of the marbles, what happens? Look at Figure 3–2c. Four marbles are stacked on top of each other; we know their porosity is 47.6%. If we add smaller marbles that will fit in the spaces between the big marbles, we will reduce the original porosity. How much we reduce it depends on how many different sizes of smaller marbles we use. If we use only size A marbles, we reduce the porosity very little because there is room for only one size—A—right in the middle of the arrangement. If we use both A and B sizes, we can add four Bs and reduce the pore space more. If we use sizes A–E marbles, we can practically eliminate the porosity. Note that we get the greatest reduction in porosity with a wide assortment of marble sizes. From this we can say that, in general, poorly sorted sandstones have low porosities.

What happens if we replace the round marbles with bricks (Fig. 3–2d)? The only pore spaces are in the cracks between the stacked bricks, so the porosity is very, very low. We can learn from the brick illustration what kind of porosity we get with square grains. But first let's digress a bit.

If we chip off all the sharp edges from a cube, eventually we end up with a sphere. Nature, too, usually begins by breaking off a more-or-less rectangular piece of rock and then proceeds to turn it into a sphere or a marble by rolling, grinding, and wearing off the sharp edges. A freshly broken piece of rock is angular; a very weathered fragment is well rounded. Between these stages, the rock passes through subangular and subrounded stages.

The more angular the sand grains, the more tightly packed they will be and the lower the porosity. The more rounded the grains, the higher the porosity.

Usually well-rounded sands are also well sorted, so they have very high porosities and permeabilities. (Permeability is the ability of a formation to allow fluid to flow. High permeability results in high production, low permeability in low production—other things being equal.) These sands are described as mature, well weathered, or even beach sand (if the grains are large). On the other hand, poorly sorted, angular, and subangular sands are said to be immature and slightly weathered with low porosities and low permeabilities.

Other Types of Porosity

Figure 3–2d also illustrates *fracture porosity*. If the bricks are rather loosely stacked, there will be spaces between them. These spaces are more in the nature of fracture planes than pore spaces. The total pore volume of a fracture system is usually very low—often 1–2%. Fractures occur naturally in rocks because of the way the earth moves and buckles over time. Although fractures have a very low porosity, they frequently have a very high permeability; large quantities of fluid can flow very easily.

Another type of porosity is *vugular porosity*, present in carbonates such as limestones and dolomites. A *vug* is a large, irregular void in the rock, usually caused when a mineral, such as calcite, is dissolved by water moving through the rock. Vugs allow large quantities of fluid to flow very easily. Caverns are examples of huge vugs.

A particular formation could have all three porosity systems or only one. Sandstones normally have only the first type of porosity, called *matrix* or *intergranular porosity*. Carbonates often have all three porosity systems: matrix, fracture, and vugular.

FORMATION ANALYSIS

Figure 3–3 illustrates some important concepts in formation analysis. Figure 3–3a depicts a unit volume of formation. This unit volume measures one unit (I) per side and has a volume of 1 cubic unit (a unit can be 1 foot, 1 inch, 1 meter—it really doesn't matter). We let the unit volume equal 100%. This means that if the unit volume is full of water, then we have 100% water. On the other hand, if the unit volume is completely filled by the matrix, then the porosity is zero and there is no water, oil, or gas present. Seldom is the porosity zero. As we saw in the previous section, if sand grains form the matrix or structure of the rock, the porosity will be well above zero in most cases.

Generally, the formation water surrounds the sand grains because of the preferential wetting of the formation by water. The water that wets the sand grains is called immovable or irreducible water. The amount of irreducible water is a function of grain size, permeability, and capillary pressure. Any excess water may be produced along with the oil or gas. The oil, or gas, is in the pore space separated from the individual grains by the formation water, both irreducible and movable. This particular unit volume in Figure 3–3 has only irreducible water and oil in the pore space. Note that the pore space is 100% full of fluid, either oil or water.

Now look at Figure 3–3b. Here the unit volume has been separated into its constituents. All of the sand grains have been compacted into the bottom and are called the matrix, the mineral from which the rock structure is made. In the case of a sandstone, the matrix is predominantly quartz. The water volume is the amount of water contained in the pore spaces of the formation. It includes both irreducible (immovable) water and free (movable) water. Above the water volume is the oil volume. Since we have a unit volume and all of the compacted matrix is at the bottom, whatever is left over is the porosity. The pore space is filled with water and oil, so they are shown in the area labeled "Porosity."

If we add up the volumes, we can say

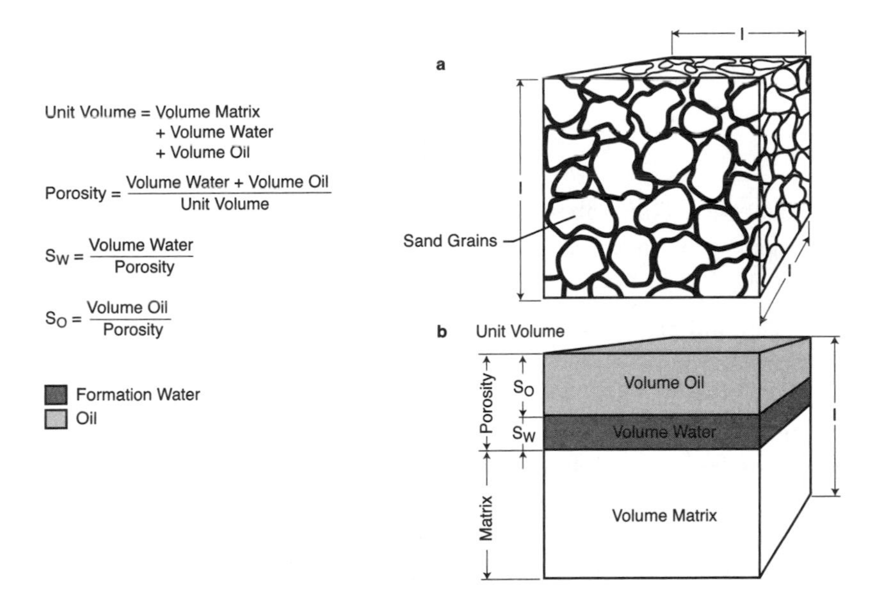

Fig. 3–3. *Unit volume and bulk volume. (a) Example of how fluids and pores are distributed in a formation. (b) The same formation with fluids separated for easy analysis. Also shown are the common equations for determining the percentage of oil and water.*

BVM	+	BVW	+	BVO	= 100% or 1.0
Bulk volume matrix		Bulk volume water		Bulk volume oil	

These volumes are called *bulk volumes.* In the figure, the bulk volume of the matrix (BVM) is 0.70 (70%); the bulk volume of the water (BVW) is 0.10 (10%); and the bulk volume of the oil (BVO) is 0.20 (20%).

We know that all of the oil and water is contained in the pore spaces; so if we add BVW and BVO, we find the porosity, 30% or 0.3. We can say

$$\text{Porosity} = \text{BVW} + \text{BVO} = 100\% - \text{BVM}$$

Although the concept of bulk volume is very useful, traditionally we have used *water saturation,* S_w, and *oil saturation,* S_o, to account for the liquids that occupy the pore space. To determine these saturations, we calculate the following:

$$S_w = \text{BVW/Porosity}$$
$$S_o = \text{BVO/Porosity}$$

Notice that $S_w + S_o = 1$, or 100%. This means the pore space is 100% full.

Note: If free gas is present, its saturation is noted by S_g. In a formation with oil, gas, and water, $S_o + S_g + S_w = 100\%$.

SHALY FORMATIONS

So far we've talked about shale-free or clean formations. But quite often, formations contain varying amounts of shale, which complicates things immensely.

Shales and clays (the terms are used almost interchangeably, although there are technical differences) are very fine-grained, platelike minerals. Because of this platelike structure and small grain size, they have immense surface areas compared to the same volume of sand grains. This large surface area causes large quantities of water to be bound to the shale mineral structures. Their fine grain size and strong capillary forces hold the water in place, so shales have essentially zero permeability and, usually, high porosity. The shale can occur as fine laminations in the sand, or it can be dispersed throughout the formation. If dispersed, the shale acts almost like another liquid in the pore space: it lowers the porosity that is available to hold fluids and reduces the permeability. The porosity available to hold fluids is called the *effective porosity*, as opposed to the *total porosity*. The total porosity includes the porosity of the shale, which is filled with bound water.

We now have a revised unit volume (Fig. 3–4): the shale matrix is included with the formation matrix (the shale and sandstone matrices are very similar except for grain size), and the shale porosity is included in the porosity section. We can determine the effective porosity by subtracting the shale bound-water volume from the total porosity. The effective water saturation is then determined by dividing the bulk volume of the formation water by the effective porosity.

To recap what we've learned so far:

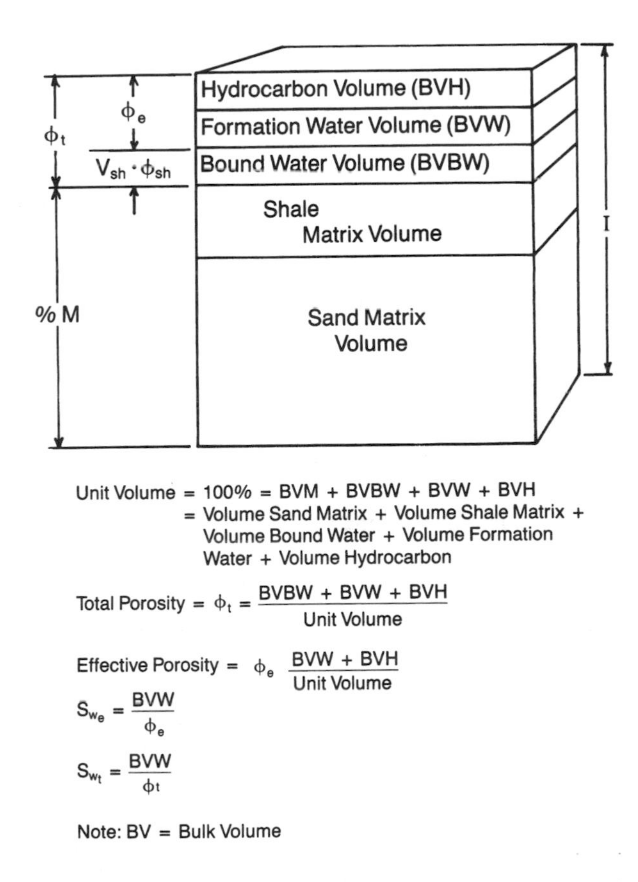

Fig. 3–4. *Unit volume and bulk volume with shale added. Note how the addition of shale changes the relationships in Figure 3–3.*

- Unit volume is a volume of formation, one unit on a side.
- Total porosity is the percent of the unit volume filled by fluids (water, oil, or gas).
- Effective porosity is the percent of the unit volume filled by fluids minus the shale bound water.
- Total water saturation (Swt) is the amount of formation water divided by the total porosity.
- Effective water saturation (Swe) is the amount of formation water divided by the effective porosity.

RESERVES ESTIMATE

From Figure 3–4 we see that we can calculate the amount of oil or gas in place in the formation if we know the effective porosity, the oil saturation, the formation thickness, and the area that the reservoir covers or that the well is capable of draining. To calculate the *reserves* (the amount of recoverable oil or gas), all we need to know in addition is a *recovery factor* (oil recovery factors are usually around 40% but may be much higher or lower). We arrive at the following equation:

Oil Reserves (N_p) = Effective Porosity (Φ_e) x Oil Saturation (S_o) x Formation Thickness (h) x Drainage Area (A) x Recovery Factor (rf)

From this equation we can see that we can recover more oil if

- Effective porosity is higher
- Oil saturation is higher (water saturation is lower)
- Formation is thicker
- Drainage area is larger (bigger reservoir)
- Recovery factor is higher

We can only estimate some of these quantities, such as recovery factor (use a best-case/worst-case scenario) or drainage area (estimated from spacing requirements, seismic information, leased area, or a best guess). The other quantities—porosity, oil saturation, and thickness—are measured by the logs.

With a reserve number, we can perform economic evaluations to determine if the well will pay out.

INVASION

So far we have been looking at the reservoir in its undisturbed state. However, drilling can profoundly affect the characteristics of a formation. The drill bit changes the rock somewhat, but the main alterations are caused by the drilling mud.

Drilling mud is a complex liquid usually composed mainly of water and suspended solids (weighting materials) as well as various chemicals that control the mud's properties (viscosity, fluid loss, acidity). Mud carries the cuttings out of the hole and up to the surface. Clays are added to water to give body to the drilling fluid. This combination makes a better carrying agent than plain water.

Another important use of the mud is to control formation pressure. Weighting materials such as barite are added to the mud so that the *hydrostatic* (or fluid) *pressure* of the mud column is greater than the formation pressure. This excess pressure stops the well from flowing or *kicking* during drilling operations. If a high-pressure zone is encountered during drilling and if formation pressure exceeds hydrostatic pressure, the driller must "weight-up" (add weighting materials) until the well is under control and pressure is balanced again.

If we take a sample of the mud and place it in a mud press, we can separate the mud into its two main components: mud filtrate and mud cake. *Mud filtrate* is a clear fluid whose salinity varies according to the source of the drilling water (the water used to make the mud) and the additives. Usually the filtrate salinity is lower than the formation water salinity. Since the filtrate is a clear fluid (no suspended solids), it can invade the formation if the pressure in the wellbore is greater than the formation pressure and can displace some of the original fluids.

The solid component of the mud is called the *mud cake*. The mud cake seals off the formation from invasion by the mud filtrate. The presence of mud cake can be detected by some logging tools. It indicates *invasion* and, indirectly, permeability. Since the mud cake is a solid, it will not normally invade the formation. Drilling mud usually cannot invade the formation because it contains a lot of suspended solids. Whole mud can be lost into the formation by inadvertently fracturing the formation or by natural fractures or vugs. In the case of fracturing, the formation strength is exceeded by the hydrostatic mud pressure, and a fracture large enough to take the mud solids is formed. This is commonly called *lost circulation* or a lost circulation zone, and large amounts of mud can be lost in a short time.

We've separated the mud into two components: filtrate and mud cake. Let's see what happens when these two components come in contact with the formation.

On the left side of Figure 3–5 is a portion of undisturbed formation. The formation we are interested in is the section of sandstone bounded above and below by shales. Note that the bottom of the sand has a water saturation, S_w, of 100%, and the upper part of the sand is at *irreducible water saturation*, S_{wirr}. (Only the irreducible water remains; oil fills the rest of the pores.) The water fills the lower portion of the formation because oil is less dense or lighter than water and rises. A transition zone exists between the upper and lower sections in which the water saturation is changing from 100% to irreducible. (Not all zones containing hydrocarbons are at irreducible water saturation.)

On the right side of Figure 3–5 is the same formation after it has been penetrated by the bit. Here, invasion has occurred. Because of the higher

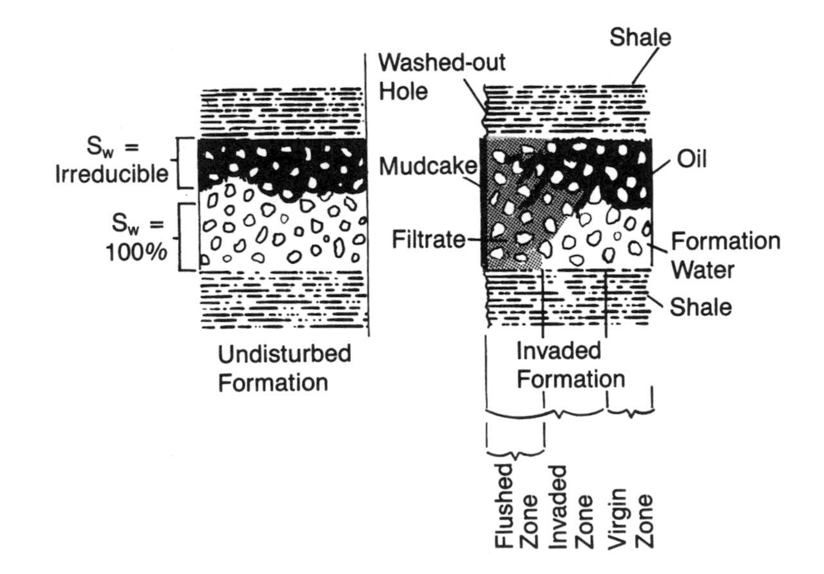

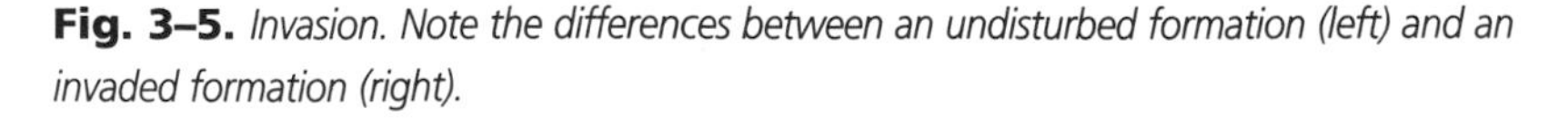
Fig. 3–5. *Invasion. Note the differences between an undisturbed formation (left) and an invaded formation (right).*

hydrostatic pressure of the mud column, the permeable formation acts like a mud press and separates the mud into mud cake and filtrate. The mud cake is formed by pressing the liquid out of the mud. The solid particles are left behind in the borehole and stick to the permeable sections. Note on the right side of Figure 3–5 that the mud cake has a definite thickness and extends from the formation. The mud filtrate, the liquid part of the mud, invades the formation. In order to invade the formation, the formation fluids that were originally there must be displaced. The filtrate flushes or displaces these fluids deeper into the reservoir and takes their place near the wellbore.

In the bottom part of the formation where S_w = 100%, the flushing is nearly complete. Most of the formation water can be moved because it is being displaced by water that is different only in the amount of dissolved salts. (In fact, sometimes the filtrate and the formation water are nearly the same salinity. In that case, it is impossible to tell whether the formation has been invaded.) The salinity of any formation water left behind will soon reach equilibrium with the filtrate because of ion exchange.

In the upper part of the zone, we have a large oil saturation ($S_w = S_{wirr}$). Although most of the formation water has been displaced by the mud filtrate, some residual oil still remains in the zone near the wellbore that was flushed by mud filtrate.

Residual oil saturation is similar to irreducible water saturation: it cannot be moved by normal means. The residual oil saturation may be abbreviated S_{or} or ROS. The residual gas saturation is designated S_{gr}. The term S_{hr}, residual hydrocarbon saturation, may be used for either gas or oil. If a formation has never contained any hydrocarbons, it will have an S_{hr} of 0. Once the formation has contained oil or gas, even if it was only migrating through the formation (that is, once S_w < 100%), the formation will have S_{hr} > 0.

Look again at Figure 3–5. The filtrate has flushed out all the original fluids possible for a certain distance from the wellbore; this volume of the formation is called the *flushed zone*, or S_{xo}. The diameter of the flushed zone is d_{xo}, the resistivity of the water in the flushed zone is R_{mf} (resistivity of mud filtrate), and the resistivity of the flushed zone formation is R_{xo}. If we go a little deeper into the formation, we find a mixture of formation fluids and mud filtrate. This zone, from the borehole wall to the end of the mud filtrate, is called the *invaded zone* and has a subscript of *i*. The diameter of invasion is d_i, the water saturation is S_i, and the water resistivity is R_i. (Resistivity is covered in detail in a few pages.) Since the water in the invaded zone is a mixture of formation water and filtrate, it is usually impossible to come up with a value for R_i. Finally, as we pass the invaded zone, we return to the undisturbed or uncontaminated formation. This is the *virgin zone*; here, the conditions are the same as on the left side of the figure.

Figure 3–6 is the same formation presented in a different fashion. On the left is a plot of S_w vs. vertical depth. At the bottom, S_w = 100% and is constant for 30 ft. or so. We then enter the transition zone, where S_w changes with depth until it reaches an irreducible value of about 25%. S_w is constant for the last 20 ft. at $S_w = S_{wirr}$ = 25%. Note that S_h, the hydrocarbon saturation, is equal to 1 – S_w.

On the right side of Figure 3–6 are three sections, drawn horizontally through the formation so that we can see how S_w varies with distance from

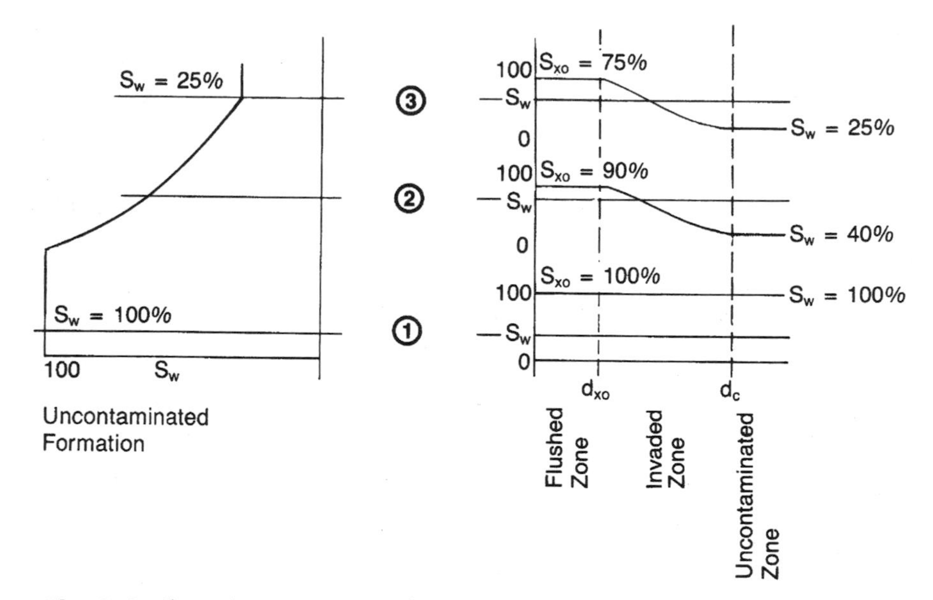

Fig. 3–6. *Change in water saturation for Figure 3–5. Saturation varies with distance from the borehole.*

the borehole at three different points in the formation. Look at the bottom section first. This represents the S_w = 100% formation. Note that S_w = 100% throughout the section, i.e., $S_{xo} = S_i = S_w$ = 100. This is because there was never any oil or gas in this part of the formation. Since we haven't added any hydrocarbons, all of the water saturations must read 100% and S_h must read zero.

Now look at the middle section. This was taken in the transition zone, where S_w was about 40%. Now we see a change in the various water saturations because oil is present in the formation, and some of it has been flushed out by invading drilling fluids. Because of the residual oil, $S_{xo} < 100\%$, $S_i < S_{xo}$, and of course S_w in the uncontaminated zone is 40%.

In the upper section, S_w is at its irreducible saturation value. Here we see a maximum variation in the various saturations: $S_{xo} < S_{xo}$ in the transition zone, and $(1 - S_{xo})$ is close to the residual oil saturation, S_{or}. The water saturation will vary throughout the invaded zone between d_{xo} and d_i. Beyond d_i, $S_w = S_{wirr} = 25\%$. In this upper section we can clearly see the flushed-zone diameter, where S_{xo} is constant, and the end of the invaded zone, where S_w becomes constant.

Be sure to study Figures 3–5 and 3–6 thoroughly. Many of the dilemmas in interpretation and evaluation arise from uncertainties about invasion.

RESISTIVITY

As we pointed out earlier, to calculate reserves, we need to measure (1) porosity, (2) the percentage of formation fluids that are hydrocarbons (S_o or S_g), (3) formation thickness (h), (4) recovery factor, and (5) area of the reservoir. Some of this data, such as formation thickness, can be obtained easily from logs; other items, such as reservoir area, must be estimated from seismic data and offset wells. Recovery factor is another item that can only be estimated initially. Log data indicate only whether hydrocarbons are present in a given formation; years of data from a particular reservoir are needed to determine the ultimate recovery factor. Log evaluation concerns itself with evaluating the only parameters we can measure: porosity, water saturation, and formation thickness. From this information we can infer production potential. But remember: these quantities are measured indirectly and never directly.

Direct vs. Indirect Measurements

Let's talk for a minute about measurements. We need to know many things during the day, and most of these can be measured: our weight, our shoe size, the time, the number of gallons of gasoline we pump into our cars. We almost take for granted our ability to make these measurements.

We assume that when the butcher weighs a pound of meat, we are getting a true pound. But how do we know that? Where does the butcher keep his pounds? Actually, he is measuring weight indirectly. He places the package of meat on the scale platform and then notes how far a spring stretches. The stretch of the spring is proportional to the weight of the object. If the scale has been calibrated, he can tell how much the object weighs in some unit, such as pounds or grams. But he does not weigh the meat directly; he weighs it indirectly by stretching a spring.

The same is true in engineering measurements. Seldom is it possible or practical to measure something directly, so indirect methods must be used.

Water Saturation

Porosity is measured in the laboratory in a manner similar to the experiment with the marbles. Unfortunately, porosity cannot be measured in the same manner *in situ* (in place in the wellbore). Therefore, various indirect methods must be used. Most of these methods use either sonic energy (the response of the formation to a sound wave passing through it) or some form of induced or applied radiation. (See Chapter 6 for more details.)

The other quantity we need to determine besides porosity is water saturation. We measure water saturation because we can't measure oil saturation. We really want to know how much oil or gas is present, not how much water. When logging measurements were first invented, the only measurement that could be made was resistivity. The measurement of porosity was still to come. In fact, the meaning of the resistivity measurement was not understood. A lot of research time and many dollars were directed at the subject. H.G. Doll and especially George Archie were pioneers in the field.

What is the tie between saturation, porosity, and resistivity? The *Archie equation* spells it out: $S_w = (FR_w/R_t)^{1/2}$. Although neither water nor hydrocarbon saturations can be measured directly in the wellbore, it is possible to infer the water saturation fairly easily by measuring the resistivity of the formation.

Resistivity is a property of a substance similar to specific gravity. It is related to electrical resistance by length and cross-sectional area. Resistance determines the amount of voltage necessary to cause a particular current to flow. Resistivity is measured in ohm-meters^2/meter. This is simplified to ohm-meters, or *ohms. Conductivity* is the opposite of resistivity (conductivity is measured in *mhos*, which is *ohm* spelled backwards). Copper has low resistivity; electric current meets little resistance in copper objects such as power lines and flows easily. On the other hand, glass has very high resistivity and is often used to make insulators. Copper has very high conductivity; glass has very low conductivity. The resistivities/conductivities found in the earth's formations generally fall somewhere between those of copper and glass.

Since sandstones are essentially glass (silicon oxide), it would seem that sandstones have very high resistivity. That is true to a certain extent. If you have a block of completely dry sandstone and measure its resistivity, the reading will be high—approaching infinity. But we know from looking at logs that sandstones often have resistivities of < 1 ohm—a very low value. So how do we resolve the differences in the resistivity between a dry sandstone and one that we find downhole? The difference is the formation water, which usually has low resistivity. We know formation water is present because the sandstone was probably deposited originally in a saltwater environment. As long as the formation has any porosity at all, there will be water within the formation.

If we measure the resistivity of distilled water, we find it has extremely high resistivity. So why does formation water usually have low resistivity? Could it be the dissolved salts? Yes. The dissolved salts that are generally found in formation water in varying amounts lower the resistivity of the water. When we measure the resistivity of our block of sandstone, its resistivity will vary with the amount of water and the salinity of the water.

Let's devise an apparatus to measure the resistivity of a block of formation; we'll call it an R-meter (Fig. 3–7). Start off with a piece of completely dry sandstone—nothing in it except sand grains and air—and measure its resistivity. The resistivity should be close to infinity. Now take some distilled water, saturate the dry sandstone block with it, and measure the resistivity again. Once more it will be close to infinity. Finally, make up a saltwater solution, measure its resistivity R_w, then flush out all of the fresh water with the saltwater solution. If we measure the resistivity of the block now, we'll see that it is much lower than before. Record these readings: R_w, the saltwater resistivity; and R_o, the resistivity of the water-saturated block (S_w = 100%).

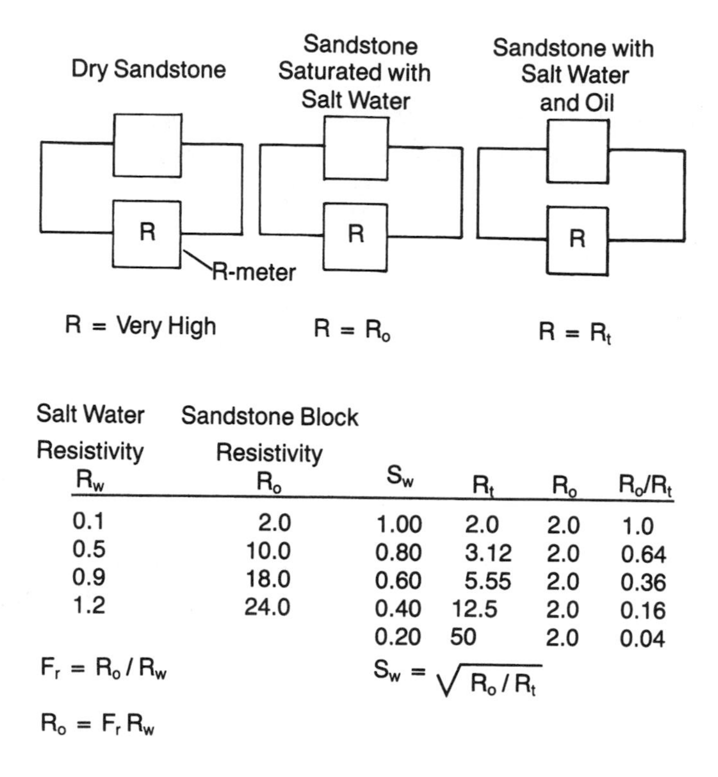

Salt Water Resistivity R_w	Sandstone Block Resistivity R_o	S_w	R_t	R_o	R_o/R_t
0.1	2.0	1.00	2.0	2.0	1.0
0.5	10.0	0.80	3.12	2.0	0.64
0.9	18.0	0.60	5.55	2.0	0.36
1.2	24.0	0.40	12.5	2.0	0.16
		0.20	50	2.0	0.04

$F_r = R_o / R_w$ $S_w = \sqrt{R_o / R_t}$

$R_o = F_r R_w$

Fig. 3–7. *Hypothetical R-meter experiment. By comparing different values of water resistivity (fresh and saline), we can deduce the formation resistivity factor, F_r.*

Now redo the experiment with a different water resistivity; this will give us a different value for the resistivity of the water-saturated block. If we repeat the experiment many times, we will see a relationship between R_w and R_o. This relationship is called the formation resistivity factor, or F_r; it is expressed mathematically as $F_r = R_o/R_w$.

If you think about what we're doing when we determine F_r, you'll conclude that the formation resistivity factor is related to porosity in some way. Not only the salinity but also the amount of water in the rock must have a bearing on R_o. What affects the amount of water that the rock can hold? The porosity. The relationship between porosity and formation factor is written $F_r = K_R/\Phi^m$, where K is a constant, usually between 0.8 and 1, and m is usually between 1.3 and 2.5. (For many years, this was the only relationship between porosity and log measurements.)

If we change our experiment and add some oil to the block, changing S_w, we can come up with a new relationship. First measure R_o by saturating the rock with saltwater. Next reduce S_w to, say, 80% by adding the oil and measure the resistivity of the rock. We will call this reading R_t for true resistivity. Adding oil increases the resistivity of the block so that $R_t > R_o$. If we reduce S_w again to 60%, measure R_t again, then reduce S_w again and measure R_t each time, using many samples of different kinds of rock, we will come up with the saturation equation, also called the Archie equation.

These are essentially the experiments that George Archie, who pioneered studies in resistivity, carried out in the early days of log interpretation research. The Father of Log Interpretation ran experiments in which he first measured the resistivity of a 100% water-saturated core and then measured the resistivity as the core was progressively saturated with oil. Archie determined that water saturation is equal to the square root of the 100% water-wet resistivity, R_o, divided by the formation (or true) resistivity, R_t:

$$S_w = \sqrt{R_o / R_t}$$

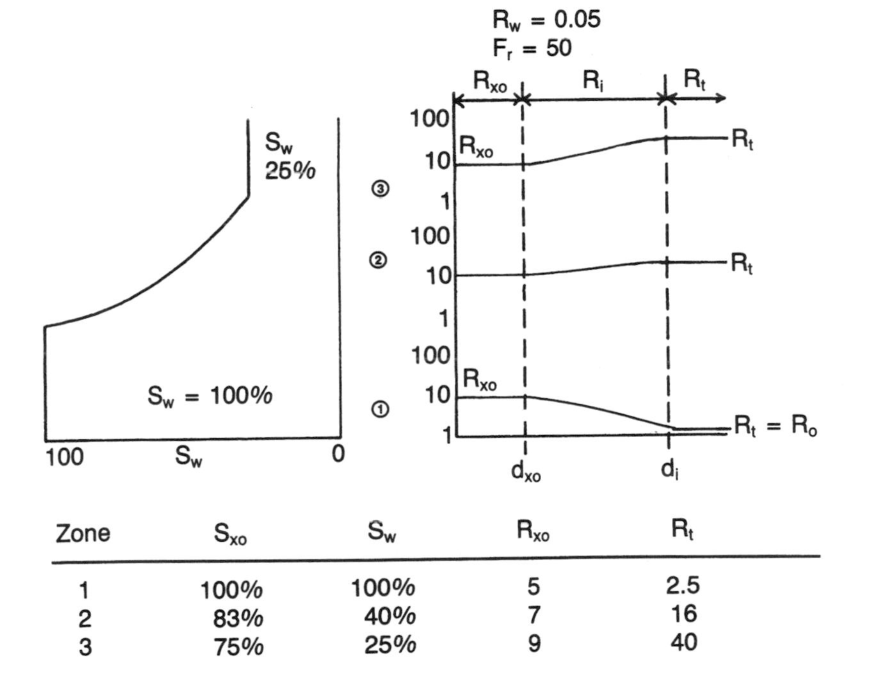

Zone	S_{xo}	S_w	R_{xo}	R_t
1	100%	100%	5	2.5
2	83%	40%	7	16
3	75%	25%	9	40

Fig. 3–8. *Resistivity profile. The figure indicates invasion by mud filtrate. Compare this with the profile in Figure 3–6.*

Now that we've talked about resistivity, let's take one more look at invasion and see what the resistivities would look like in different parts of the formation.

The presence of invasion by the mud filtrate gives rise to a resistivity profile, as shown in Figure 3–8. Here, the water saturation varies from 100% in the bottom of the zone through a transition zone to $S_w = 25\%$ at the top. Notice the three resistivity profiles on the right side of the figure. On all of the profiles, the flushed-zone resistivity, R_{xo}, next to the borehole is very high. It is slightly higher in the oil-bearing zones, because of the residual hydrocarbons, than in the 100% wet zone.

In the uninvaded or virgin part of the formation, there is a large contrast in resistivity between the flushed zone and where $S_w = 100\%$. The contrast is not nearly as great through the transition zone, and there is very little contrast in resistivities in the low water saturation interval.

The resistivity close to the wellbore is generally higher than the true resistivity of the formation, even when the formation has hydrocarbons. (It is possible to have $R_{xo} < R_t$ when drilling with very low-resistivity mud systems or when the water saturation is very low.) Since the depth of invasion is unknown, tools have been developed that measure the resistivities at various depths around the wellbore (see Chapter 5). Before we advance to that topic, though, we need to study one of the most basic logs of all: the mud log.

4

Mud Logging

As the drill bit cuts through the different formations, the cuttings are brought to the surface by the drilling mud. Traces of oil or gas may also be brought up in the mud. The practice of mud logging tries to identify, record, and/or evaluate lithology, drilling parameters, and hydrocarbon shows. The information obtained by the mud logger is presented in the form of various logs such as the driller's log, the cuttings log, or the show evaluation log. The mud logger takes this information, correlates it with data from other wells, and determines whether the well may be able to produce hydrocarbons in commercial quantities. In addition, the mud logger monitors the wellbore for stability to prevent blowouts or kicks, and he makes sure information is relayed to the right people at the right time.

To comprehend all of the information available, we need to understand four important areas of mud logging: rate of penetration and lag, gas detection, formation evaluation and sample collection, and show evaluation.

RATE OF PENETRATION AND LAG

Rate of penetration (ROP) is the oldest and most common way of measuring and evaluating formation characteristics and drilling efficiency. The formation's lithology (rock type and hardness), porosity, and pressure affect the ROP. The drilling parameters that affect ROP include the weight on the bit, the bit's speed (rpm), the drillstring configuration, the type of bit selected and its condition, and hydraulics.

Measuring ROP

Mud loggers measure ROP manually in three main ways: strapping the kelly, observing the drilling rate curve, or checking Geolograph™ charts. When the rig crew straps the kelly, they mark it in some increment of depth (feet or meters) and then record how much time is needed to drill each interval. When using the drilling time curve (Fig. 4–1), the logger observes a circular or strip chart, graduated in units of time, which moves a pen calibrated in units of depth. He notes the time taken to drill a certain number of feet or meters and can then determine ROP. The Geolograph chart, a third means of measuring ROP, is a strip chart on a drum that rotates once every 24 hours. Like the drilling time curve chart, this chart is also marked in intervals of time. Here, though, each increment of length is recorded as a tick mark.

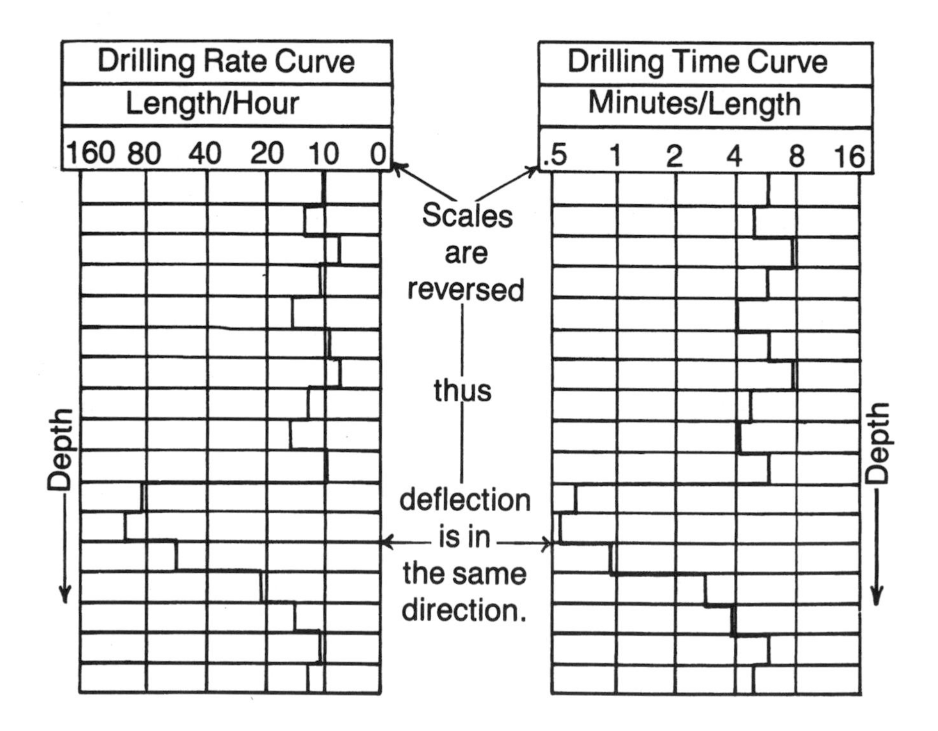

Fig. 4–1. *Drilling time curve. Note that units can be either in length per hour or in minutes per length, yet the curves are essentially identical (adapted from Anadrill's* Delta Manual, *4–4).*

When rate of penetration is presented, it is a plot of ROP vs. depth. Usually it is included on mud logs along with the parameters for correlation and interpretation. ROP can also be plotted automatically with an online plotter.

If ROP is expressed in units of length/hr, its curve is called a drilling rate curve. When the units are in min/length, the curve is referred to as a drilling time curve.

Interpreting ROP from the Mud Log

Figure 4–2 is a simplified ROP curve. On it are several lettered areas that illustrate the kinds of information we can read from the log.

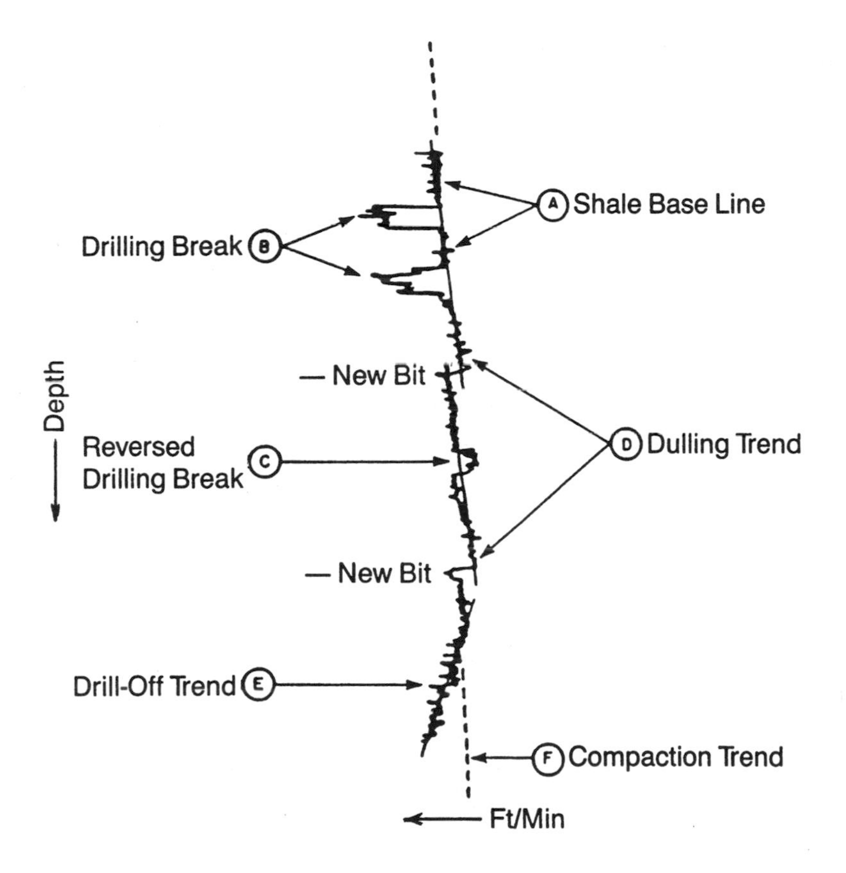

Fig. 4–2. *ROP curve terminology (adapted from Anadrill's* Delta Manual *4–7).*

The *base line* (A) is a reference point for interpretations; it simplifies correlations. Usually the rock chosen for the base line is one of the hardest lithologies that the bit will drill through. In this formation, we use shale as the base line; however, limestone is often used as the base line in carbonate sequences. Whatever the lithology, the base line establishes a norm.

Any deflections from the norm signal changes in the formation rock. In this example, the deflections are interpreted as a sand/shale sequence. The *drilling break* (B) is the deflection. It usually indicates a change in lithology, although it sometimes is the result of crossing a fault. Whatever the case, the drilling break notes an abrupt increase in the ROP—usually two or more times greater than the baseline average.

Occasionally we note a *reversed drilling break* (C) on the track. This indicates an abrupt decrease in the ROP and can imply changes in lithology. Reverse drilling breaks usually are associated with very dense formations called *caps*. They may also denote a shale/sand interface or a formation where production has depleted the formation's pressure.

As the bit wears out, it drills less efficiently. The ROP curve shows this change as a slope away from the baseline (D). The *dulling trend* can help the driller know when a bit needs to be changed.

A *drill-off trend* (E) is a gradual, usually uniform increase in the ROP. It often indicates a transition zone where pore pressures are increasing.

As the overburden pressure and age of the rocks increase with depth, the formation becomes more compacted. This *compaction trend* (F) can sometimes be seen on the log over long intervals. See Figure 4–3 for the drilling responses of some common rock types.

Lag

Lag is the amount of time that elapses from the moment when the bit penetrates a new formation until the moment when the downhole particles and/or traces of gas travel back up the wellbore to the surface. ROP is measured instantly; as soon as the bit increases or decreases its speed, the driller has input. However, samples from that particular formation may not be circulated up for several minutes. The mud logger must keep this in mind as he correlates the data.

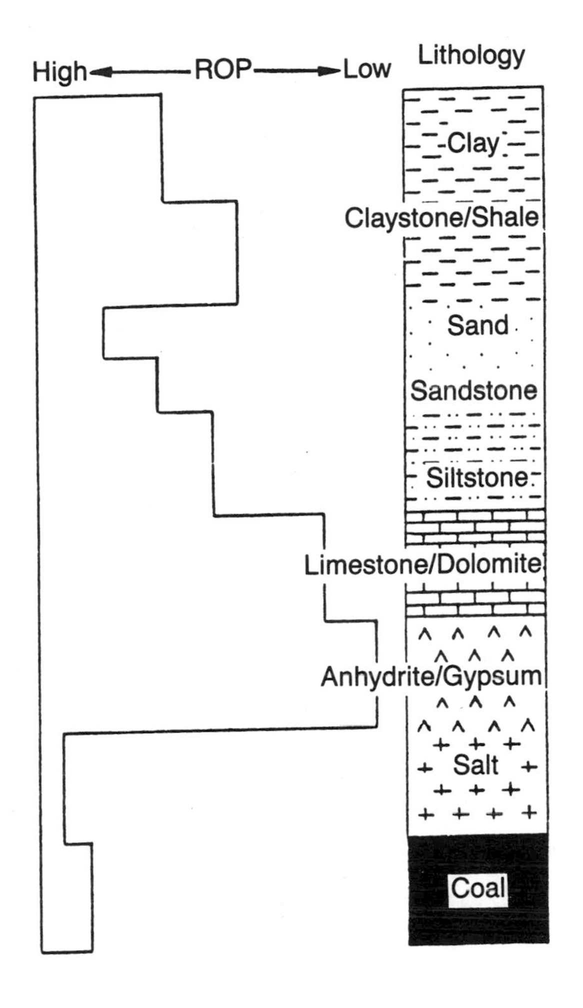

Fig. 4–3. *Drilling responses of common rock types (adapted from Anadrill's* Delta Manual, *9–3).*

GAS DETECTION

Gases extracted from the mud system are usually the first indication that hydrocarbons are present downhole. Gas enters the drilling fluid from one of three sources: (1) a gas-bearing formation, (2) a formation feeding gas into the mud, or (3) contamination.

As the bit drills through a formation, it opens or exposes some of the pores. Fluid from these opened pores mixes with the drilling mud. This gas, along with cuttings, or pieces, from the drilled formation, is pumped back up toward the surface. As the gas and cuttings rise, the pressure drops and more gas comes out of the pores in the cuttings. This "liberated gas" is an important piece of data for log interpretations.

If the hydrostatic pressure is less than the formation pressure, even more gas can flow into the wellbore. The amount of flow from the formation into the borehole depends upon the *pressure differential* (the difference between the hydrostatic pressure and the formation pressure), the porosity and permeability, the properties of the formation's fluids, and the length of time this condition lasts. When the formation fluids enter continuously, the well is said to *kick*. *Swabbing* (lifting the drillstring rapidly) also encourages formation fluids to flow into the well because the wellbore pressure drops. Engineers can identify this gas that enters the borehole during swabbing, called *connection gas*, and can use the data to enhance formation evaluation and improve well safety.

Occasionally gas is introduced into the drilling fluid from a source other than the formation, particularly when oil-based drilling fluids are used. This is called *contamination gas*. However, this case is rare.

Gas Detectors

Gas in mud is measured when the drilling fluid returns to the surface. A collector or trap is used to remove gas samples from the return line. The trap samples the mud returns consistently and reliably, regardless of the circulation system's flow rate. In addition to continuous sampling, random batch samples are collected, and the gas components are extracted.

Gas can be detected in five main ways: thermal catalytic combustion [TCC, or hot-wire detector (HWD)], gas chromatography (GC), thermal conductivity detector (TCD), flame ionization detector (FID), or infrared analyzer (IRA).

TCC instruments, more commonly known as hot-wire detectors, have been around for a long time because they are simple and inexpensive and they perform adequately. However, the instruments are unstable, so responses vary, and the method fails at gas concentrations that exceed a few percent. Nevertheless, this technique remains primary, especially when supplemented by other procedures such as gas chromatography.

Gas chromatography results are more accurate and more quantitative than hot-wire methods; however, they take minutes rather than seconds to complete. Therefore, TCC is used to detect the presence of hydrocarbons in the mud returns, while GC is used to analyze the composition of the gas stream on a regular but intermittent basis—usually after a show is detected with TCC.

The *thermal conductivity* detector is the least sensitive device normally used for monitoring hydrocarbons. Under optimum conditions, the detection limit of hydrocarbon in air is about 1%. However, the TCD has good linearity (uniform response over a wide range of measurements), is easy to use, and is durable and inexpensive.

Flame ionization detectors are popular for gas analysis instrumentation used outside of the mud logging industry. They are superior in many ways to other systems. However, the FID is expensive and difficult to operate, which limits its use.

The final device, the *infrared analyzer*, may be operated continuously but only for one compound at a time. In addition, IRAs cost more than TCC devices and are sensitive to vibration and power supply variations. Otherwise, they are easy to operate and have a sensitivity comparable to TCC methods.

A combination of two or more of these five methods helps mud loggers detect the presence of gas and analyze its components.

Analyzing Returns

The entire gas detection and analysis process is very orderly. The sequence yields data on the content of gas in mud, oil in mud, gas in cuttings, and oil in cuttings. These data constitute the *hydrocarbon log*, a continuous record organized by well depth.

The continuous gas sample is usually analyzed by the TCC or HWD gas detector. These devices are not calibrated to an absolute scale, so any gas responses or *shows* are relative. The results are usually reported in units, such as a 200-unit show.

After a show is detected with the TCC, a sample is analyzed with the gas chromatograph. This device reports the analysis of the gas as percentages of methane (C_1), ethane (C_2), propane (C_3), butanes (C_4), and pentanes (C_5). In gas associated with an oil show, there will be a higher percentage of C_3, C_4, and C_5. In a gas show, C_1 and C_2 will be predominant.

If the gas saturation of the oil is low, an oil show may not be detected by the gas-in-mud response. In these cases, loggers try to detect the presence of oil in the mud in other ways. Sometimes oil can be identified simply by visual examination. Since the mud is largely water, the oil will float on the surface of the water and may be detected by color, sheen, or oil globules on the surface of the mud pit or the samples. Color, intensity, and fluorescence (oil fluoresces under a black light) are then noted.

Gas may also be detected in fresh cuttings. A sample of cuttings is pulverized in water within a sealed container. After pulverization, the vapor above the cuttings-fluid level is analyzed with a hot-wire detector or gas chromatograph. The logger also looks at the sample to detect any shows of oil, such as sheen or fluorescence.

Measuring and Recording the Readings

Total mud gas readings are recorded continuously with a recorder chart. The gas readings are recorded for each logging interval and represent the gas response vs. depth. Since there is a lag between the time when a formation is drilled and the time when the sample reaches the surface, gas readings must be lagged or adjusted for the time it takes for the gas to reach the surface after the formation is drilled.

In logging total gas readings, several measurements are usually designated separately.

- **Total gas reading**—The maximum reading during a specific interval or the total meter reading at any point.
- **Background gas reading**—Either *drilling background gas*, which is the average gas reading while drilling in low-permeability zones such as shale, or *circulating background gas*, the average reading when circulating the bit off bottom.
- **Connection gas reading**—The difference between the drilling background gas and the total gas reading that occurs during a drillpipe connection.
- **Trip gas reading**—The maximum total gas reading from bottoms up after a trip. During this time, which may be several hours, gas can accumulate in the mud at the bottom of the hole. When mud circulation resumes, the mud that was sitting on the bottom of the hole is called the *bottoms*. When that mud is circulated to the surface, you have "bottoms up." Often, a kick will show up on the monitors; this kick is called trip gas.

A data pad (Fig. 4–4) is used to record several kinds of gas data. Columns A–G are fairly common on all forms. Column A records when a connection is made (a new section of drillpipe is added), and column B records the interval drilled. In column C, the drilling rate is determined for each interval. The stroke counter records the number of strokes made by the reciprocating mud pump. The end strokes reading, column D, records the end of each interval from the stroke counter. A lag stroke counter, which is set a certain number of strokes (the lag) behind the stroke counter, is also used. When the lag stroke counter equals the end stroke counter, the interval just drilled will be at the surface. Column E, the total mud gas reading, records the reading for each depth interval when the lag counter reading equals the end stroke reading in column D.

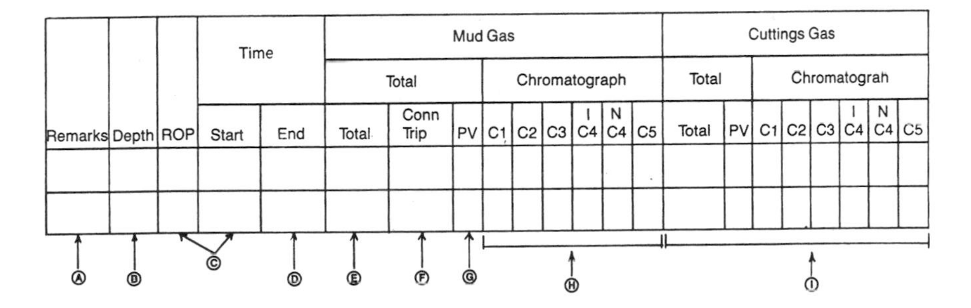

Fig. 4–4. *Simple data pad. Note the discussion in the text for the values of each column (adapted from Anadrill's* Delta Manual, *7–16).*

Gas peaks that occur from connections, trips (when the entire drillstring is pulled from the hole, as in changing a bit), surveys (measurements such as hole deviation taken on the drillpipe), and other down time is designated in column F. Let's say we make a connection at 8043 ft. We observe an abnormal gas peak of 90 units during the lagged interval 8040–8045 (the time required to circulate cuttings from the bit to the surface). This gas peak is due to the *connection gas*—gas that seeps out from the formation during down time. The value of the connection gas is the total gas reading minus the background gas.

Column G is for reading the vapors of the thermal catalytic combustion filament. The value represents the presence of heavier hydrocarbons in the sample. In some areas, this type of detection is required in addition to chromatography, which is recorded in column H.

The final column (I) is used in areas where the cuttings are collected, broken up, and measured. The gas content readings are called *cuttings gas readings* and are used primarily to interpret permeability.

The total gas readings from column E are presented on the recorder chart as a continuous curve that records data immediately. They are also plotted on the formation analysis log, the show analysis log, and/or the pressure evaluation log. The most common correlation curve is the ROP curve; it is usually presented on any log on linear or nonlinear scales with bar graphs or point-to-point plots (Fig. 4–5).

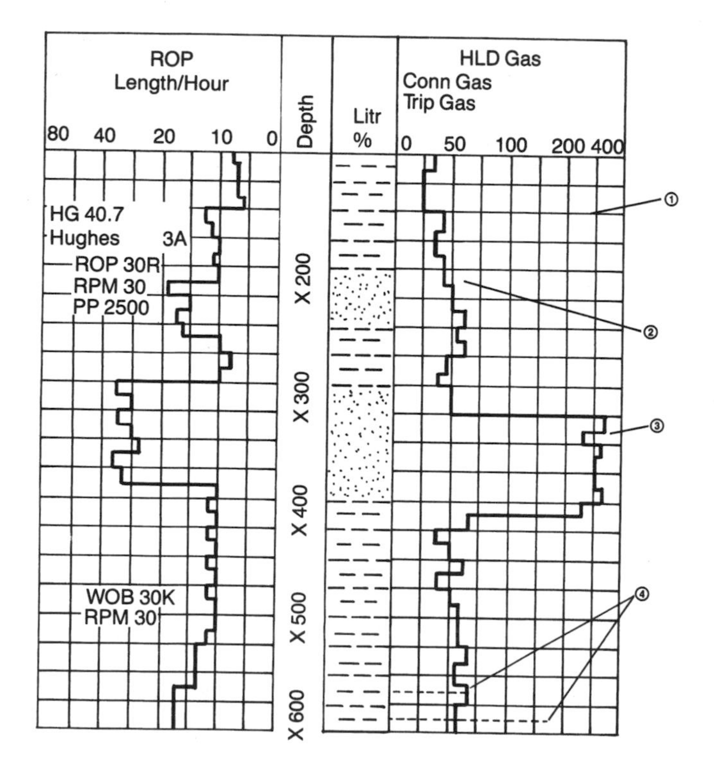

Fig. 4–5. *Gas curve presentation. The ROP curve can be used with other curves as a correlation curve. The curves indicate (1) trip gas, (2) drilling background, (3) gas show, and (4) connection gas (adapted from Anadrill's* Delta Manual, *7–18).*

Total gas measurement can be applied in three ways. First, it evaluates hydrocarbon shows. If the reading increases, hydrocarbons are present in a zone. The reading itself does not indicate productivity; however, increased readings in a potentially permeable, high-porosity zone often indicate the zone may be productive. Second, gas measure-ment detects pressure. Increasing background gas usually indicates increasing formation pressures. Connection gas and abnormally high trip gas usually indicate a nearly balanced mud system (hydrostatic pressure = formation pressure) or even an underbalanced system. Finally, the total gas reading curve can be correlated with other measurements, such as resistivity, ROP, spontaneous potential, and offset well curves. These correlations can provide input on potentially productive zones.

Interpretations

When the drill bit penetrates a hydrocarbon-bearing formation, only a small amount of hydrocarbon enters the mud and is mixed with large quantities of circulating fluid. The size of show depends on drilling and sampling factors unrelated to the amount of oil or gas in the reservoir. In general, we can conclude the following.

- The amount of gas in mud may be misleading in evaluating the quality of a reservoir. A formation that is highly prospective (good saturation, porosity, and thickness) will yield relatively small amounts of gas in mud if the *gas/oil ratio* (GOR) is low. A less prospective formation where the GOR is high may produce large gas-in-mud indications. A prospective formation, which is drilled very slowly with a higher rate of mud circulation than was used previously, may yield a smaller gas-in-mud indication than a less prospective interval that is drilled rapidly.

- Complete flushing can result in no oil or gas in the mud. Vuggy or fractured reservoirs saturated with oil that has a low GOR and residing at great depths where drilling is slow represent a worst case for successful mud logging.

- The uncertainty in evaluating flushing, retaining saturation, and losses in the return line reduce the possibility of using a numerical classification successfully in interpreting a reservoir.

COLLECTING SAMPLES

One of the most important jobs of the mud logger is to collect a representative sample of drill cuttings from the shale shakers and prepare it for lithological identification and hydrocarbon show evaluation.

When the cuttings arrive at the shale shaker, they are covered in mud, unsorted by size, and generally unidentifiable. The shale shaker sifts or separates the larger cuttings from the drilling fluids and *fines*—microscopic or dust-sized pieces of formation. The fluid is filtered for reuse, and the cuttings are routed to the reserves pit. The mud logger collects some of the cuttings before they are routed to the reserve pit and lost.

Once the samples are collected, the logger can examine them unwashed and wet, washed and wet, or washed and dried. The unwashed samples are collected directly from the shale shakers and are untreated. They are placed in labeled sample bags and are shipped to a laboratory. Washed samples are also collected from the shale shaker; however, the excess drilling mud is flushed away and then the samples are sieved to remove the coarser *cavings* (formation fragments that originate from the sides, not the bottom, of the borehole) before they are put in sample bags. Washed and dried samples go through the same steps as washed samples. Before they are bagged, though, they are air dried or dried in ovens. It is from this set of samples that cuttings are taken for microscopic analysis for lithology identification and to observe oil shows.

Care must be taken to obtain representative samples, not only the cuttings that arrive at the shaker last. The exception is spot samples, which are taken to locate precisely the depth of a particular formation top. Once the top is measured, the balance of the samples should be collected on the shale shaker screen and a representative sample bagged. The bagged samples are sent either to the oil company or to a laboratory for analysis.

Sample Description

The mud logger is responsible for describing the samples. Some of the more commonly encountered rock types are described in Table 4–1.

Argillaceous Rocks

Clay—Complex, platy aluminosilicates < 2 microns in size. Two basic types recognized are expandable (clays that swell upon contact with water, such as montmorillonites) and nonexpandable (illites).

Claystone/Shale—Same mineral content and size as clays but indurated by compaction and dewatering. In cuttings, it is difficult to distinguish between the two. Shale must break into platelike particles.

Marl—Any clay rock (from clay to shale) with 35–65% calcareous content.

Arenaceous Rocks

Siltstone—Clay-based rock in silt-sized grains or quartz particles. Any rock or intermediate composition between clay-based and sand-based rock.

Sand/Sandstone—Pure sand grains or sand grains with a clay matrix. Grain sizes are fine to very coarse and angular to rounded. Grains poorly to well sorted, cementation poor to good.

Carbonates

Limestone—Primarily calcium carbonate, recognized by fizzing strongly with 10% HCl. Some appear granular, but there are a number of classifications.

Dolomite—Similar to limestone but with a substantial part of the calcium replaced by magnesium. Less fizzing than limestone.

Evaporites

Anhydrite—Calcium sulfate or gypsum. White when pure; usually soft.

Halite (Rock Salt)—Sodium chloride. Can occur in large domes or in layers. Soft and soluble in water.

Carbonaceous Rocks

Coal—Black or dark brown, vitreous carbon. May be hard and brittle. Also occurs as peats, lignites, and other forms of organic matter.

Accessory Minerals

Pyrite—Iron sulfide. A light brassy yellow mineral associated with all sedimentary rocks. Its hardness and chemical stability may cause drilling problems if it occurs in large quantities.

Glauconite—Dark green to black iron silicate related to the mica group.

Mica—Calcium, magnesium, and iron silicate; platy in appearance.

Table 4–1. *Sample Descriptions*

SHOW EVALUATION

A show is the presence of hydrocarbons in a sample over and above background levels. Show evaluation is the complete analysis of the hydrocarbon-bearing formation with respect to lithology and type of hydrocarbon present. A complete show evaluation identifies the presence and type of hydrocarbon, determines the depth and thickness of the show, assesses the porosity and permeability, and assigns a show value that indicates the potential productivity of the formation.

Two types of shows are recognized: gas and oil. A gas show is hard to identify, but the mud logger may see a significant increase in gas levels. An oil show is an increase in heavier-than-methane gas levels as well as a physical indication of oil.

Identification

Four tests are used to detect hydrocarbons: odor, staining, fluorescence, and cut.

Odor does not usually apply to cuttings, but it is a useful test on cores. Although difficult to standardize, odor falls into one of four categories: poor, slight, fair, or strong.

Oil *staining*, like odor, is more useful when applied to cores. In general, the more desirable oils are light to colorless, while viscous oils are dark. Staining is described in terms of both color and percentage of sample stained, e.g., patchy, laminated.

Liquid hydrocarbons *fluoresce* under ultraviolet light, and the amount, intensity, and color are the first and best indications of a show. Intensity is subdivided into none, poor, fair, or strong; color is subdivided into the values in Table 4–2.

API Gravity	Fluorescent color
< 15	Brown to none
15-25	Orange to none
25-35	Yellow to cream
35-45	White
> 45	Blue white/violet

Table 4–2. *API values of fluorescence*

Cut defines the leaching of oil from a sample by a solvent. Trichloroethane is the most common agent. A cut is made by taking a sample of fluorescing cuttings and adding a few drops of solvent. The oil leaches out of the sample, and the fluorescence passes from the sample into the solvent. The rate at which the oil leaches out is classed as *flash, streaming* (instant, fast, slow), or *crush cut.* Intensity and color are also recorded. Finally, the residue that is left around the side of the spot dish is noted in white light. A dark brown, non-fluorescent ring implies bitumen, or *dead oil*, rather than any producible hydrocarbons.

The type of show, whether oil or gas, also needs to be determined. A gas show can be identified by an increase in total gas or by a gas chromatograph analysis showing gases heavier than methane. An oil show can be identified by one or all of these: (1) visible signs of oil on the surface of the mud, (2) fluorescence, or (3) increase of heavy hydrocarbon gases. In addition, depth and thickness help establish a formation's productivity on the simple premise that the thicker the show, the larger the volume of oil that can be extracted from the reservoir.

Porosity

In addition to these methods, porosity, permeability, and hydrocarbon ratio assist in evaluation. Porosity can be determined on the basis of ROP measurements. The faster the rate of penetration, the more porous the rock. This evaluation can help determine the relative porosities over the extent of the show. The first indication that a porous rock has been drilled is a drilling break. Mud loggers can also look at the samples under a microscope to make a visible estimate of porosity.

Permeability

Permeability measurements require special equipment that cannot be taken to the field, so samples must be used in the lab. The ease with which the oil is leached from the sample is an indication of its permeability. As mentioned earlier, a flash cut means the oil was leached rapidly and implies good permeability. Streaming cut denotes moderate permeability; crush cut signifies poor permeability.

Hydrocarbon Ratio Analysis

Hydrocarbon ratio analysis relates the quantities of methane, ethane, propane, butane, and pentane to reservoir fluids (gas, oil, water). If the hydrocarbon ratios are plotted for each sample through a show, the gas/oil and oil/water boundaries may be established.

Application

Complete show evaluation can help us (1) identify the presence of hydrocarbons and (2) make recommendations for coring and testing programs. In coring, the standard procedure in oil companies is to circulate up a drilling break and analyze the mud and rock for signs of gas or oil. On the basis of this analysis, a core may be cut—sometimes cut continually until the oil/water contact is passed. After this, drilling may continue. In addition to coring, show evaluations correlated with offset wells and wireline logs can assist in reservoir interpretations.

The mud log continues to be one of the first indicators of producing formations. In Figure 4–6, note the show at 9582 to 9590. A lot of information is reported, from a formation perspective—fine grained, friable, glauconitic, subrounded, slightly calcareous sandstone with a trace of kaolinitic clay—to the good gas show on the right side of the log. The drilling time curve correlates fairly well with other wireline logging parameters. While the mud log is a very useful tool, it serves more as an indicator of when and where to look more closely at the formation. Starting with Chapter 5, we'll look at the wireline logs that can confirm the information gleaned from the mud log.

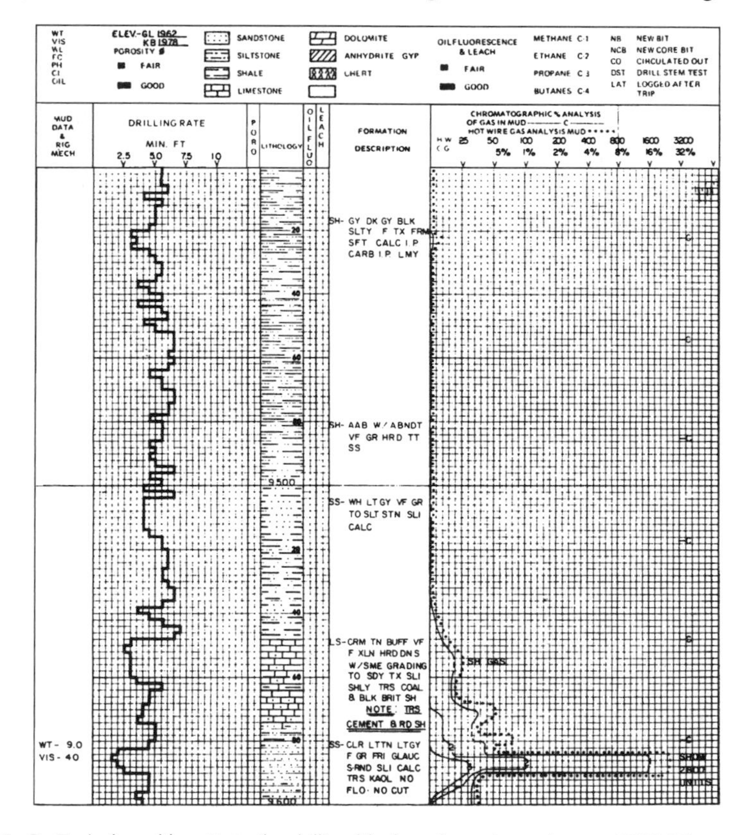

Fig. 4–6. *Typical mud log. Note the drilling block and good gas show at 9582-90. See text for details.*

5

Resistivity Measurement

As we saw in Chapter 3, different formations have different resistivities. More importantly, formation resistivity varies with porosity, water salinity, and hydrocarbon content. Although we cannot directly measure the amount of hydrocarbon in a formation, we can infer or estimate the volume of oil or gas with the aid of resistivity measurements.

Three types of logging tools are used to measure formation resistivity: *induction logs*, *focused resistivity logs*, and *unfocused resistivity logs*. (Note that the word "log" is used interchangeably for both the tool and the curve.) These tools can be further divided into those that measure a very small volume of the formation—*microresistivity logs*—and those that measure a relatively large volume of the formation.

A primitive form of the unfocused resistivity log was the first log run on an electric wireline. (Electric wireline is a wire rope or cable with insulated electrical wires or conductors beneath the strands of cable.) This device was invented and developed by two French brothers, Conrad and Marcel Schlumberger.

As the logging industry grew, new resistivity tools were introduced that gave more accurate readings, were easier to interpret, or worked in different environments than the original electric log. Today, the most common resistivity tools are the dual induction-focused log, the dual laterolog-microfocused log, the microlaterolog, and the microlog.

Almost from the beginning, engineers realized that more than one resistivity measurement was needed because of the effects of invasion. In Figure 5–1, the water saturation for the virgin formation is shown on the left side. At the bottom of the interval, the zone is 100% wet (S_w = 100%). The resistivity of the flushed zone, R_{xo}, is 5 ohms, while the resistivity of the uninvaded zone, R_o, is 2.5 ohms. (Remember from Chapter 3 that $R_o = R_t$ when S_w = 100%.)

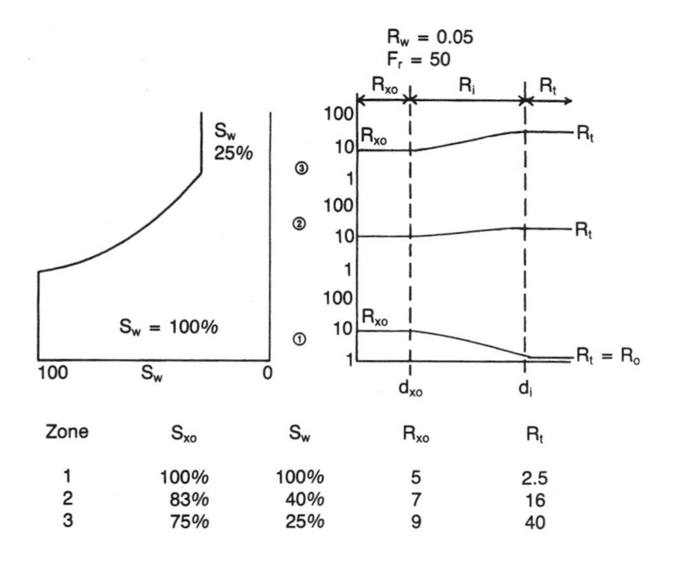

Zone	S_{xo}	S_w	R_{xo}	R_t
1	100%	100%	5	2.5
2	83%	40%	7	16
3	75%	25%	9	40

Fig. 5–1. *Resistivity profile of invaded zone. Resistivity varies with distance from the borehole and with water saturation.*

R_{xo} next to the borehole is higher than R_o because the mud filtrate in the flushed zone usually has a higher resistivity than R_w in the uninvaded portion of the formation. Occasionally, though, engineers run into a formation where the mud filtrate resistivity is less than R_w. This may occur when a well is drilled with salt muds or at very shallow depths where the formation water is more likely to be fresh (less salty, higher resistivity).

At the top of the interval, S_w = 25%, R_{xo} is about 9 ohms, and the true resistivity of the formation (R_t) is about 40 ohms. The R_{xo} value is higher in the top of the zone than in the bottom because some of the oil was left behind when the mud filtrate flushed the zone, and oil and gas have high resistivities. In other words, S_{xo} < 100% but greater than S_w if the zone contains oil or gas; $R_t > R_o$ because of the hydrocarbon present. The oil and gas fill some of the pore space, so there is less room for formation water. Since the resistivity of the formation depends on the amount of the formation water present (other things, such as salinity and porosity, being equal), the resistivity must increase as the volume of water decreases.

As you can see, the effects of invasion cause the resistivity to vary close to the borehole. Sometimes it is high; at other times it is low, depending on the resistivity of the mud filtrate, the formation water, the water saturation, and the porosity. The farther the measurement is taken horizontally from the borehole—into the formation—the more nearly it will match the true resistivity of the formation.

One of the problems we encounter when using resistivity tools is that few of these tools read deeply enough to measure the formation's true resistivity (see Table 5–1). However, because we know the approximate resistivity profile in the invaded formation and because we can measure the porosity (with other logging tools) and estimate R_w, we can construct correction curves or charts that will give a good approximation of R_t. With a good R_t, we can calculate S_w. If S_w is low, the zone is potentially productive; if S_w is high, the zone is wet.

Tool	Uses	Measures	VR*	DI*	Limitations
Electric	Freshwater mud, thick beds	R_t, R_i	16 in. to 20 ft.	16 in. to 20 ft.	Difficult to interpret thin beds
Induction	Freshwater mud, air or oil mud	R_t, R_i	5 ft.	5-20 ft.	R_t, > 100, R_m, < R_w, salt muds
Dual Induction	Freshwater mud	R_t, R_i	18 in. to 5 ft.	30 in. to 20 ft.	R_t, > 100, R_m, < R_w, salt muds
Laterolog	Salt mud	R_t	12–32 in.	80 in.	
Dual Laterolog	Salt mud	R_t, R_{xo}	24 in.	> 80 in.	
Microlog	Freshwater mud	Indicates permeability, Cal,	2 in.	< 4 in.	Porosity > 15, h_{mc} < ½ in.
Microlaterolog	Salt mud Freshwater mud	R_{xo}, cal	2 in.	< 4 in.	h_{mc} < ½ in.
Microspherically Focused Log	Salt mud Freshwater mud	R_{xo}, cal	2 in.	< 2.5 in.	

Table 5–1. *Resistivity Tools: Uses and Limitations*
** VR = vertical resolution; DI = depth of investigation*

INDUCTION TOOLS

In some areas, the water-based, conductive mud damages the formations by causing water-sensitive clays to swell. Formation permeability is thereby reduced and results in serious production problems. To avoid damaging the formations, many wells are drilled with oil-based mud or with no mud at all (air-drilled wells). Unfortunately, the electric log does not work in oil-based mud or in air. It needs a mud column to conduct the current from the tool into the formation. The *induction resistivity tool* was developed to provide a means of logging wells drilled with oil-based (nonconductive) muds or with air.

Although the induction resistivity tool was developed to operate in a nonconductive mud, engineers soon recognized that the tool worked better than the original electric log in freshwater muds. The induction curve was easier to read, and it read closer to true resistivity in formations where the resistivity was not greater than 200 ohms and $R_{mf} > R_w$.

The induction tool works by using the interaction of magnetism and electricity. When a current is sent through a conductor, a magnetic field is created. If the current alternates, the magnetic field also alternates by reversing poles at the same rate at which the current is alternating. If a conductor is moved through a magnetic field, a voltage is induced in the conductor. Voltage can also be induced in a stationary conductor by alternating the magnetic field. The induction tool uses these principles in its operation.

We see this principle in action in Figure 5–2. A high-frequency *transmitter current* flows through a coil mounted in the logging *sonde* or tool. This current sets up a high-frequency magnetic field around the tool, extending into the formation. The constantly changing magnetic field causes currents to flow through the formation concentric with the axis of the induction tool. The currents, called *ground loops*, are proportional to the conductivity of the formation; they alternate at the same frequency as the magnetic field and the transmitter current flowing through the transmitter coil. The ground loop currents set up a magnetic field of their own. This secondary magnetic field causes a current to flow in the receiver coil located in the induction sonde. The amount of current flowing in the receiver coil is proportional to the ground loop currents and therefore to the conductivity of the formation. The signal in the receiver coil is detected, processed, and recorded on the log as either a conductivity measurement (C) or a resistivity measurement ($C = 1/R$).

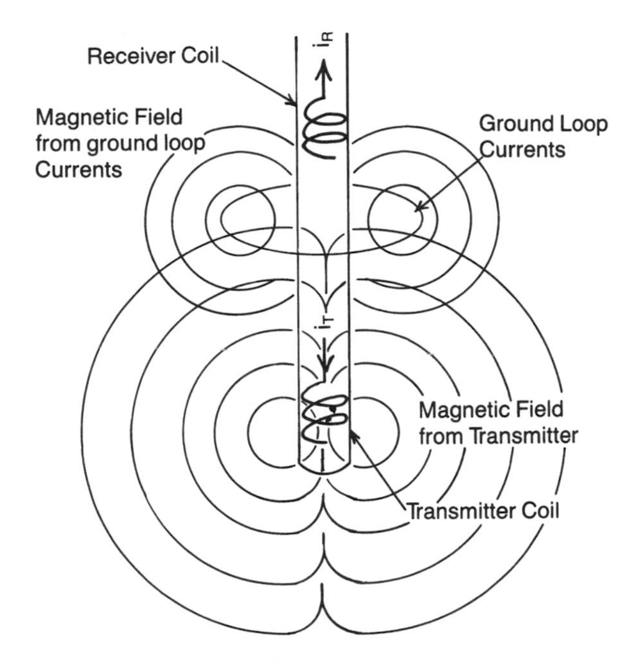

Fig. 5–2. *Schematic of an induction tool. A high-frequency transmitter current induces ground currents, which in turn generate a signal in the receiver coil.*

The tool illustrated in Figure 5–2 is a two-coil device. In practice, extra coils are used to help focus the effects of the main transmitter and receiver coils and to remove unwanted signals from the borehole (*borehole effect*) and adjacent formations (*bed boundary effect*). One popular induction tool has six different coils. The depth of investigation (the depth where most of the measurement is obtained) for a typical deep induction tool is about 10 ft. The *vertical resolution*—that is, the thinnest bed that the tool will detect—is 40 in. Both the depth of investigation and the vertical resolution are affected by the spacing between the main transmitter and the receiver coils as well as by the placement of the focusing coils. By judicious selection of these parameters, we can design different depths of investigation into a tool. Thus, we can measure the resistivity profile through the invaded zone and correct the deep induction reading to move it closer to the R_t value we want.

For many years, the *induction electric log* was the most popular induction tool in high-porosity, moderate-resistivity formations such as in California and along the Gulf Coast. A single induction curve with a vertical resolution of about 3 ft. and a depth of investigation of about 10 ft. was combined with either a short normal curve or a shallow laterolog curve. (These two resistivity curves will be described in the next sections.) Since invasion is seldom deep when porosity is high, these two curves, corrected for borehole and bed boundary effects, could be used to determine R_t.

The *dual induction laterolog* was developed for those areas that had lower porosities and deeper invasion than California–Gulf Coast-type formations. The tool has two induction curves (IL_d and IL_m) with a vertical resolution of about 40 in. However, the IL_d curve reads very deeply into the formation, while the medium-induction IL_m curve reads only about half as deep. A shallow-reading laterolog combined with the two induction curves gives a good description of the resistivity profile. Figure 5–3 is a typical *tornado chart* (so called because of its distinctive shape) used to correct the dual induction spherically focused log to obtain R_t.

The development of microprocessors and onsite computing power has led to still newer induction tools that were not feasible with the earlier technology. The concept of using multiple focused arrays of induction coils was proposed in the late 1950s. The mass of data that had to be processed made development of the concept impractical until the 1990s, when vastly increased data handling and improved tool designs were available. These new induction tools use multiple sets, or *arrays*, of induction coils.

By combining various coil spacings, operating at several frequencies simultaneously, and measuring portions of the induction signal previously unused, the *multiple-array induction tools* can measure the resistivity profile radially. The log records resistivities at 10, 20, 30, 60, and 90 in. from the borehole. Vertical resolutions of 1, 2, or 4 ft. can be selected by the logging engineer. The tool can measure resistivity in formations with a ratio of $R_t/R_m \leq 500$ and also when $R_{xo} < R_m$. The actual mud resistivity is measured at downhole conditions and used to correct the log readings for effects caused by hole rugosity (roughness) and drilling mud resistivity.

Induction tools were first used in boreholes with nonconductive fluids, but today they are used mostly in conductive, water-based drilling muds. Although the effects of the borehole have been minimized by tool design, all induction readings are affected to a certain extent by the mud-filled borehole. The effect of the borehole is best minimized by using single or dual induction tools only when formation resistivities are < 100 ohm and when $R_{mf} > R_w$. Charts correcting for the borehole signal are available from logging companies. (See Figure 5–4 for recommendations on which type of tool to use.)

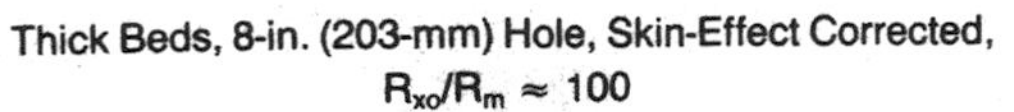

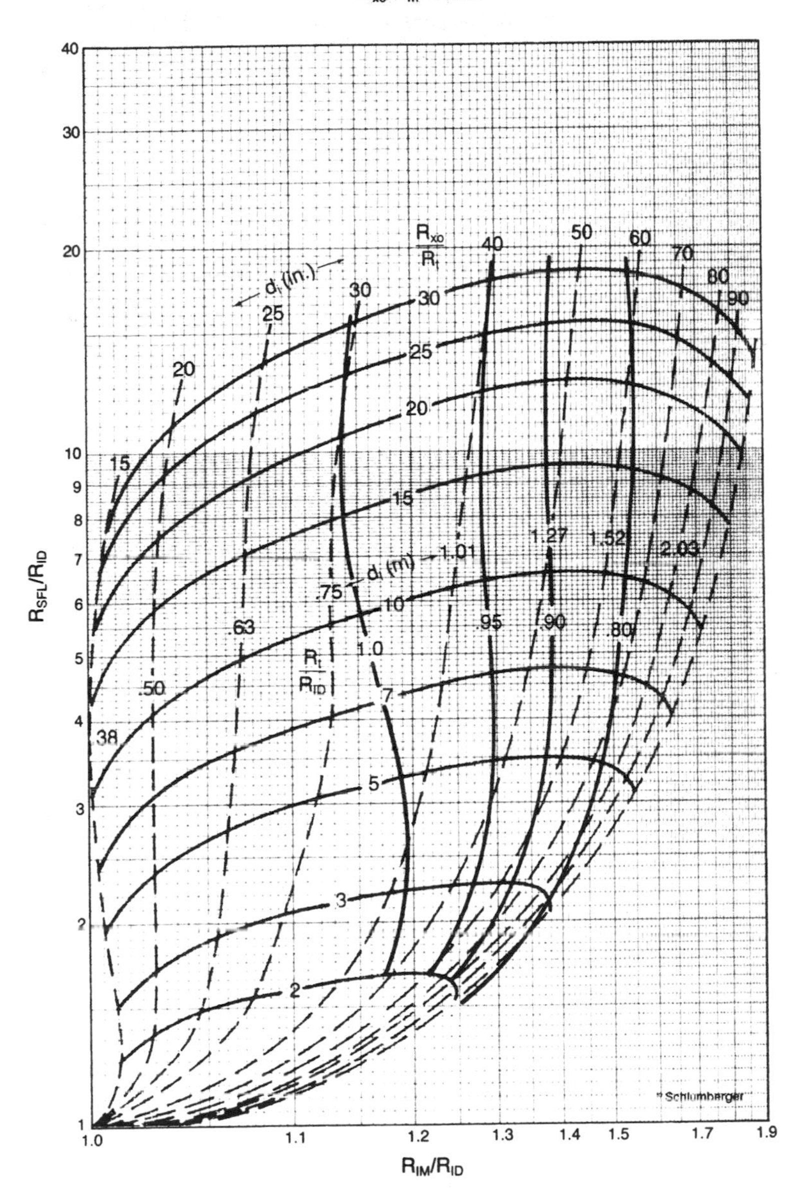

Fig. 5–3. *Tornado chart (courtesy Schlumberger). Curves are used to correct deep induction resistivity to true resistivity.*

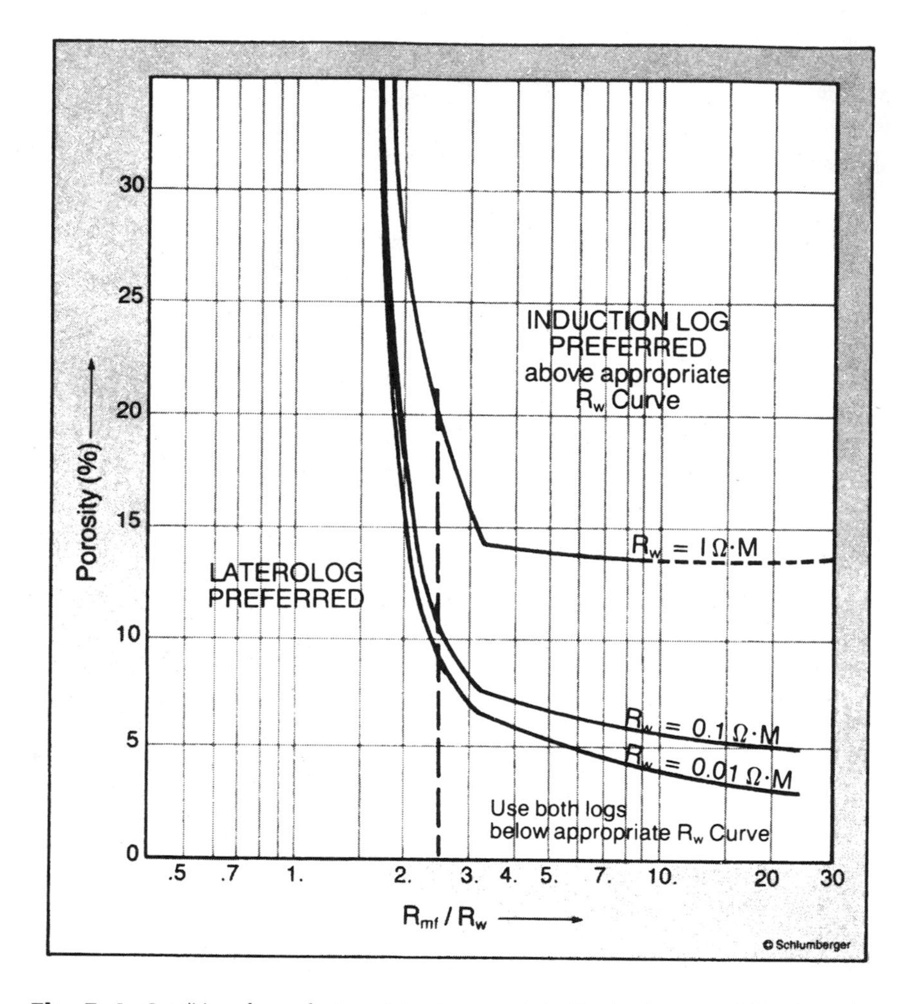

Fig. 5–4. *Conditions for preferring a laterolog or an induction log (courtesy Schlumberger).*

FOCUSED ELECTRIC LOGS

In highly resistive formations or in very low-resistivity muds, neither the induction log nor the electric log works very well. The mud column tends to short-circuit the current of the electric log, causing the readings to be too low and lacking in definition. In the case of the induction log, the mud column contributes a large portion of the total signal when the contrast or ratio between formation resistivity and the mud resistivity is high. The induction reading is also too low in these conditions. A better measuring device than either the induction log or the electric log was needed for highly resistive formations, such as those in the Midcontinent and the Rocky Mountains. Focused electric logs were developed to fill this need. These tools are generally used when the R_t/R_m ratio is high, i.e., when the formation resistivity > 100 ohms and/or $R_m < R_w$ (usually the case when salt muds are used).

With focused electric logs, the measuring current is forced out into the formation by focusing electrodes. Figure 5–5 illustrates the simplest of the focused logs, a three-electrode laterolog or guard log. Three electrodes are mounted on a sonde and are insulated from each other. The upper guard, or focusing electrode, is called A_1; the lower focusing electrode is A_2. The center electrode is the measuring electrode, A_0.

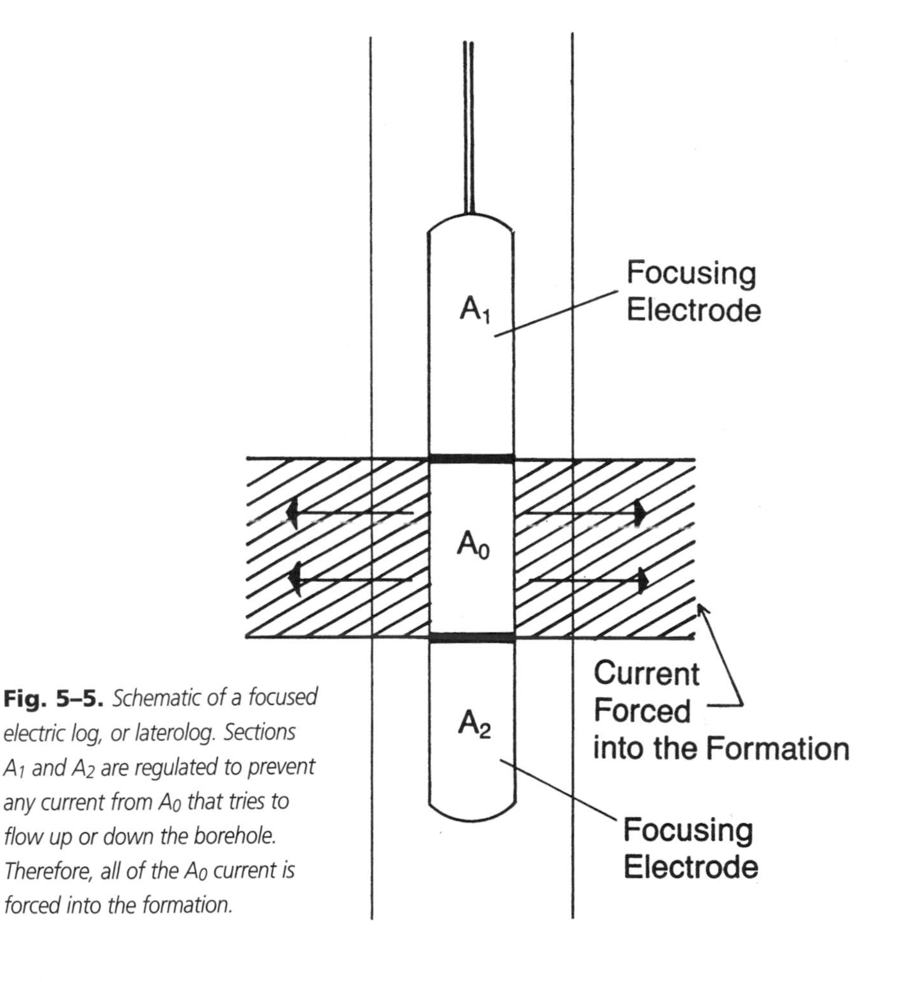

Fig. 5–5. *Schematic of a focused electric log, or laterolog. Sections A_1 and A_2 are regulated to prevent any current from A_0 that tries to flow up or down the borehole. Therefore, all of the A_0 current is forced into the formation.*

A constant current is emitted from A_0. The two focusing, or guard, electrodes are regulated so that any current that tries to pass them is focused and then forced into the formation instead. Therefore, low mud column resistivity or high formation resistivity has little effect on the measuring current, and accurate resistivity measurements are obtained.

Various laterolog tools have been developed over the years. Among the currently used tools, the *dual laterolog* is most common. This tool, similar to the dual induction log, has both a deep and a shallow-measuring laterolog curve; it is often run in conjunction with a very shallow-reading laterolog tool that is mounted on a pad pressed against the borehole. This shallow-reading curve, called the *microspherically focused log*, measures the flushed-zone resistivity. This combination of measurements can define the resistivity profile created when mud filtrate invades the formation.

Since the current path for these logs is through the mud column to the borehole wall, through the invaded zone, and then to the uncontaminated zone, the resistivity readings are a combination of these different zones. Mud and invaded zones affect the laterolog tool's resistivity measurement much less than the measurements made with unfocused tools, a feature that minimizes corrections. When corrections are needed, charts (such as Fig. 5–6) may be used to correct the resistivity readings for bed thickness, mud column, and invasion effects to derive a better estimation of R_t from charts like Figure 5–3.

To determine which tool (induction or laterolog) to use, first estimate what the formation water will be (from a nearby well, for example). Second, estimate R_{mf}. Again, information from a nearby well can be used if the mud system will be the same. Third, calculate R_{mf}/R_w. Fourth, estimate the porosity of the zone. The numbers don't have to be exact, just reasonable guesses. For example:

Well A

R_w = 0.04–0.08
R_{mf} = 0.2–0.3
ϕ = 15–25%
R_{mf}/R_w = 0.2/0.08 to 0.3/0.04 = 2.5 to 7.5

If R_{mf}/R_w = 2.5 to 7.5 and porosity is 15–25%, the induction log is preferred. If R_{mf} = 0.02 (salt mud system), R_{mf}/R_w = 0.02/0.04 to 0.02/0.08 = 0.5 to 0.25 and the laterolog is preferred.

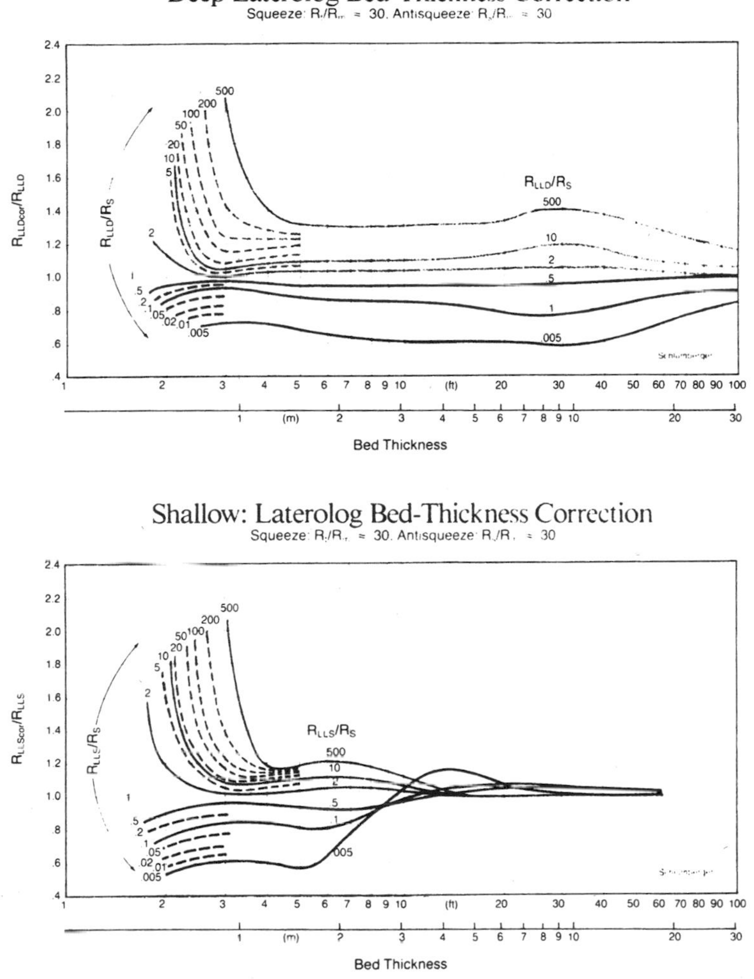

Fig. 5–6. *Typical bed-thickness correction chart (courtesy Schlumberger).*

ELECTRIC LOGS

The electric log, or *E-log,* is of interest because so many of these old logs still exist in log libraries and are used by geologists to map formations and put prospects together. Although electric logs are mechanically simple, they are very difficult to read because of borehole effects and the many effects of the various electrode arrangements.

Figure 5–7 illustrates a simple four-electrode system called a *normal device.* There are usually two normal measurements on an electric log. One has a short spacing of around 10–16 in. The other, called the long normal, has a spacing of 64 in.

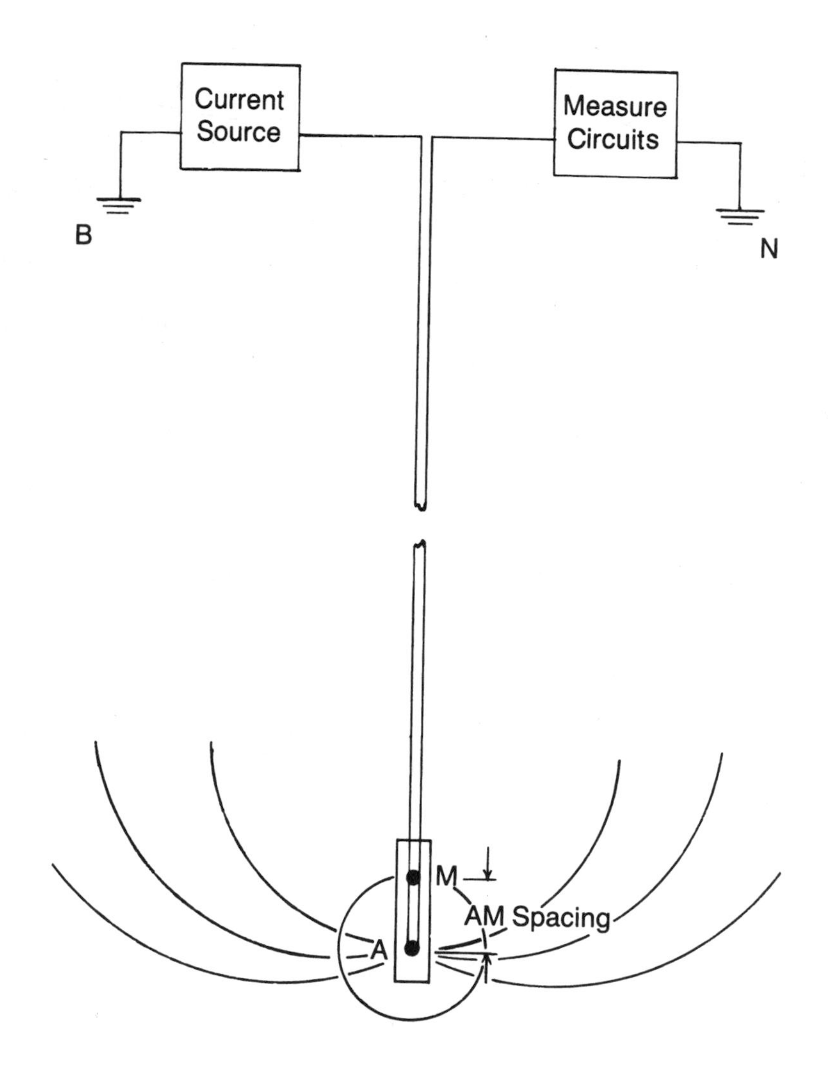

Fig. 5–7. *Schematic of electric log, normal device. Current flows from electrode A to ground electrode B. Voltage is measured by electrode M with respect to another ground electrode at N.*

Current from a constant source is emitted from electrode *A* and returns to electrode *B*, which is a long distance away. The current leaves point *A* in an essentially spherical manner. The voltage at *M* is measured with respect to a reference electrode (*N*) that is at 0 voltage. Since the current emitted by *A* is constant, any variation in the voltage at *M* will be due to changes in the resistivity of the formations (from Ohm's law, which states voltage = current X resistivity X area ÷ length).

Another four-electrode device is called the lateral, which works on the same principle as the normal. The main difference is that three electrodes are on the sonde: *A*, *M*, and *N*. Electrodes *M* and *N* are close together, and the voltage difference between them is measured. The advantage of the lateral over the normal is that it has a deeper radius of investigation (the depth to which it will read into the formation)—about 18 ft. However, lateral curves are difficult to interpret, even for experienced log analysts.

The electric log usually consists of four measurements: spontaneous potential (SP) and three resistivity measurements with different depths of investigation—16-in. short normal, 64-in. medium normal, and 18-ft., 8-in. lateral. In some parts of the United States, a 10-ft. lateral is run in place of the 64-in. normal curve.

Determining true resistivity from an electric log is more an art than a science. Many rules must be followed; many corrections and judgments must be made. The thickness of the beds (formations), the mud resistivity, the shale resistivity, and the type of measuring device—normal or lateral—must all be weighed when trying to extract an accurate resistivity from these logs. Fortunately, the old E-log has been replaced by much more easily read logs, such as the induction electric log and the laterolog.

In spite of these difficulties, do not dismiss the electric log lightly just because modern and more easily read logs are available. E-log readings often give a good indication of formation resistivity and invasion. A lot of oil and gas has been discovered with these logs.

SPONTANEOUS POTENTIAL

The spontaneous potential curve—or SP, as it is more commonly called—is generally included on resistivity logs. SP is not a resistivity measurement; rather, it is a naturally occurring voltage or *potential* (an early term for voltage) caused when the conductive drilling mud contacts the formations. Since the voltage occurs naturally, it is spontaneous.

The origin (and even the existence) of spontaneous potential was hotly debated in the early days of logging. Although not as crucial to log interpretation as it once was because of the abundance of other logging measurements, the SP log nevertheless can provide much useful information, especially on older logs.

The most common uses of the SP are to correlate between logs, to identify permeable formations, to measure zone thickness, to calculate R_w, and to identify shaliness. The SP log can be run only when the borehole is filled with a conductive mud because the mud filtrate is a crucial ingredient in generating the SP voltage. The log cannot be run in oil-based mud or in air-drilled holes.

The source of the SP is the combination of the drilling mud (especially mud filtrate), the formation water, invasion, and the presence of sand and shale. The SP voltage is caused by an electrochemical action resulting from the differences in the salinities of the various fluids. In essence, it is a wet-cell battery, similar to a car battery.

Fluids cannot flow through the shale because of the shale's very small grain size. However, shale's layered makeup and the negative electrical charges on these layers allow the sodium ions (Na^+) to flow through shale but block chlorine ions (Cl^-). If a permeable sandstone's formation water contains sodium ions and is separated by a shale from another solution (the drilling mud) that contains a different concentration of ions, current will flow as positively charged sodium ions migrate. If the drilling mud has fewer sodium ions than the formation water, current will flow from the formation water through the shale to the wellbore. This flow is possible because of *membrane potential*, the voltage caused by the shale acting as a sieve that filters out everything except positively charged sodium ions.

Membrane potential accounts for about 80% of the spontaneous potential. The remaining SP is caused by the mud column, or mud filtrate, being in contact with the formation water. Since chlorine ions (Cl^-) are more mobile than sodium ions, more negative ions than positive ions will cross the junction between the two fluids. This effect is called the *liquid junction potential*. We add it to membrane potential to find the total SP measurement or deflection.

The amount of spontaneous potential in a clean (shale-free) formation is proportional to the ratio of the mud filtrate resistivity to the formation water resistivity (R_{mf}/R_w). We can use this relationship to calculate R_w.

Track 1, located to the left of the depth column, is called the *SP track.* SP is always measured in millivolts per division. Note in Figure 5–8 that the fairly constant reading opposite the shales is called the *shale baseline.* When opposite a sandstone, the SP will normally deflect to the left (away from the depth track) if $R_{mf} > R_w$. If a sand is thick enough, it will generally reach a constant reading called the static SP (SSP), the maximum SP that could be measured if no current were flowing in the borehole. If the sand is not thick enough or the formation resistivity is high, the SP will not reach its maximum value and corrections will have to be made.

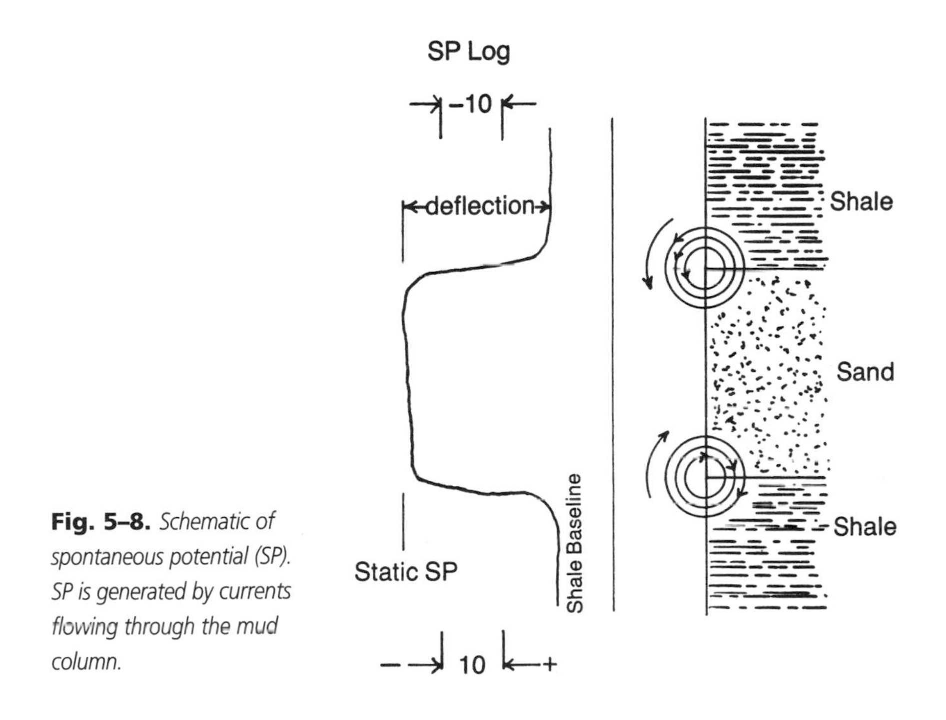

Fig. 5–8. *Schematic of spontaneous potential (SP). SP is generated by currents flowing through the mud column.*

In low-resistivity sands (for wet sands, S_w = 100%) where R_t is approximately the same as the mud resistivity (R_m), the SSP value will be reached when the bed thickness (h) is about 15 times as large as the hole diameter (d). For the normal range of hole diameters, this means a bed thickness of 7–15 ft. If h is about twice as great as d, the SP curve will reach about 90% of SSP. However, in high-resistivity formations that may be hydrocarbon bearing, a much thicker formation is necessary to develop the SSP reading. If $R_t > 20 \times Rm$, then in a bed $15 \times d$, the SP will only be about 90% of SSP. We need a zone nearly 40 ft. thick to develop 100% of SSP in these conditions. For a zone only $2 \times d$ in thickness, the SP will reach < 30% of its true value.

To have spontaneous potential, a permeable formation like a sandstone must lie next to a shale. But what happens if an impermeable, highly resistive formation like a limestone or dolomite is present? First of all, to generate spontaneous potential, there must be some permeability, and carbonates usually do not have SP because they generally have low permeability. Second, the high resistivity of the carbonates encourages the SP currents generated by the sands and shales to stay in the borehole fluids until they have passed the zone of high resistivity and can reach a sand or shale (see Fig. 5–9).

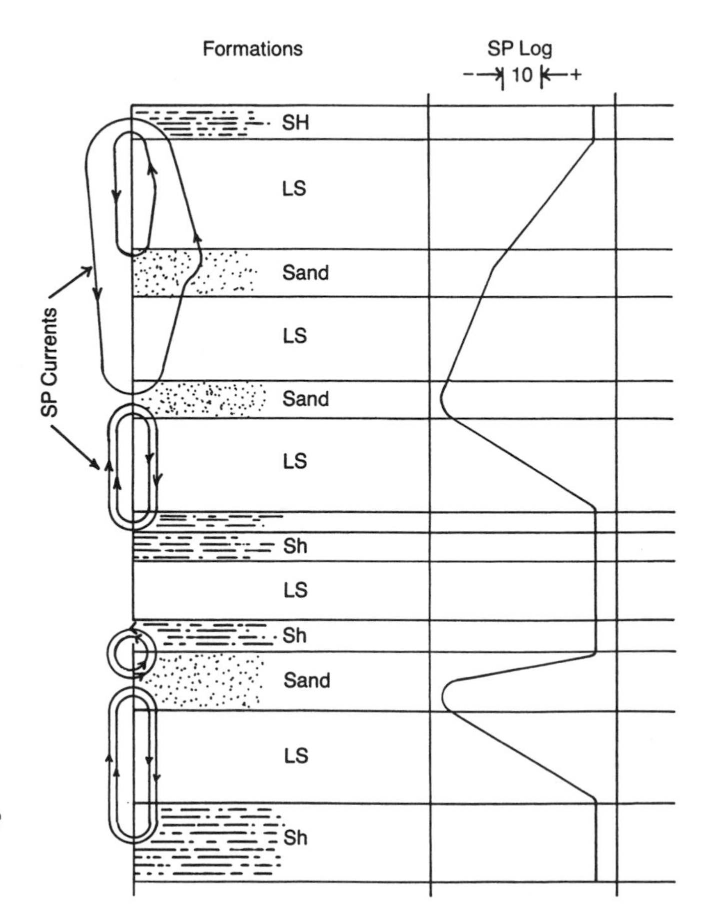

Fig. 5–9. *SP curve shapes, sand/shale/ limestone series.*

The presence of limestone greatly complicates interpretation of SP. Normally, it is easier to identify lithology from some other log, such as a porosity log. The R_w calculation is much less reliable because we cannot always be sure we are measuring the SSP. Notice on Figure 5–9 that the SP curve through the limestone sections tends to be a straight line connecting permeable zones. The limestone may make a formation (either sand or shale) appear thicker than it really is.

MICRORESISTIVITY TOOLS

Microresistivity tools are designed to read R_{xo}, the resistivity of the flushed zone. Since the flushed zone may be only 3 or 4 in. deep, R_{xo} tools are all very shallow reading. The electrodes are mounted on flexible pads pressed against the borehole wall, thereby eliminating most of the effects of the mud on the measurement.

The R_{xo} measurement has several uses. In certain techniques, water saturation may be obtained from resistivity ratios if S_{xo}, the flushed-zone water saturation, is known or can be estimated. An advantage of the resistivity ratio method is that water saturation can be calculated without using a porosity measurement. In addition, formation resistivity factor (F_r) can be calculated from the equation $F_r = R_{xo}/R_{mf}$.

Micrologs

The first microresistivity tool was the microlog. On this tool, a pad carrying electrodes is filled with an insulating oil. The pad is pressed against the wall of the hole by the backup pad. Current flows along a path, as in Figure 5–10. The two resistivity curves are closely related to the electric log's resistivity measurements in theory but are on a much smaller physical scale.

The printed log consists of a caliper curve in track 1, which is used to measure the diameter of the borehole and to indicate the presence and thickness of mud cake, and two resistivity curves—the 1-in. × 1-in. *microinverse* and the 2-in. *micronormal*—in tracks 2 and 3. The microinverse curve has a shallower depth of investigation than the micronormal curve.

The difference in depths of investigation is used to indicate permeability. If a formation has been invaded, the mud solids will build up as mud cake on the face of the formation. The resistivity of the mud cake is usually lower than that of the formation immediately adjacent to the wellbore. Since the

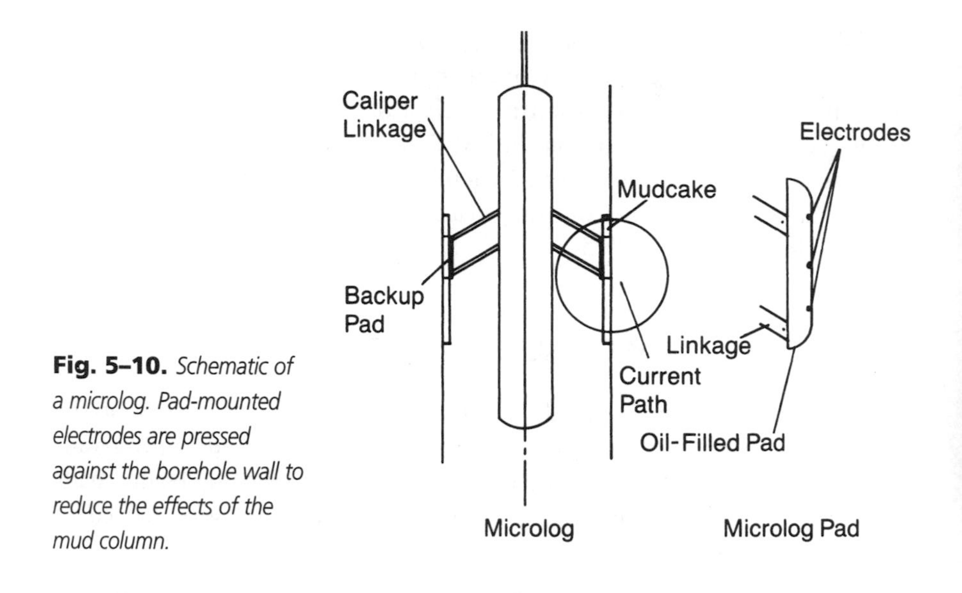

Fig. 5–10. *Schematic of a micr解log. Pad-mounted electrodes are pressed against the borehole wall to reduce the effects of the mud column.*

microinverse log makes a very shallow measurement, it will "see" primarily the mud cake. The micronormal will read more deeply into the formation and see some of the invaded zone. Therefore, the micronormal resistivity will be higher than the microinverse if the zone has been invaded. If invasion has not occurred, the two curves will read about the same.

The micrololog is best used to indicate permeability and formation thickness (because of its very sharp bed boundaries) in low- to medium-resistivity formations. However, at one time, charts were used that calculated porosity from micrololog readings on the basis of assumed relationships between porosity, formation factor, and flushed-zone water saturation (S_{xo}). This method is still used when evaluating logs run before the advent of other porosity tools. Although micrologs are still run today, they are less important because better R_{xo} devices and much better porosity devices have been developed.

Microlaterologs

The microlaterolog (MLL) is similar in operation to its big brother, the laterolog (LL). The tool carries small concentric electrodes on a flexible pad pressed against the borehole wall. The outer guard electrodes force the current into the formation and prevent short-circuiting by the mud cake. For this reason, the microlaterolog is used in high-resistivity formations. A caliper log and usually a micrololog are also recorded.

Microspherically Focused Logs

The microspherically focused log (MSFL) uses the same principle of operation as the spherically focused log but on a smaller scale. It is also a pad device and is often combined with other measurements such as the dual laterolog or the formation density log (to be discussed later). Like the MLL, the MSFL is recorded on a logarithmic scale.

Microresistivity tools are essential in evaluating potentially productive hydrocarbon formations. From these logs, we can often determine characteristics such as the following:

- depth of invasion
- flushed-zone water saturation (S_{xo})
- moved hydrocarbons ($S_{xo} - S_w$)
- corrections for deep induction and laterolog readings
- permeability
- hole diameter
- zone thickness
- porosity

In this chapter we have tied in the idea of resistivity profiles due to invasion, which we read about in Chapter 3, with methods of determining the resistivity profile. The preferred resistivity measuring tools depend on the types of formations and their resistivities. In low to medium resistivities, single or dual induction tools are used; in high resistivities, laterolog devices are best. Microresistivity tools are used to determine the resistivity of the flushed zone and the invaded zone.

We also learned about the SP curve, which is often used to correlate different logs on the same well, to determine formation water resistivity (R_w), to give a sand count (net pay thickness), and to indicate permeability. The SP curve is normally included on all resistivity logs except microresistivity logs, and it may also be included on other logs such as the sonic, which is covered in Chapter 6.

6

Porosity Measurements

In Chapter 3 we discussed porosity, the space within reservoir rocks that can contain fluids. Porosity is measured as a percentage of the bulk volume; a formation with no porosity has 0% porosity. Most reservoir rocks have porosity ranging from 6–30%; the greater the porosity, the more fluid the rock can hold. Therefore, petroleum engineers and geologists are extremely interested in porosity.

If we could directly measure the porosity of the formations as they lie in the earth (in situ), petroleum exploration would be simple. Unfortunately, porosity is one of those variables that defies easy determination. Since we cannot measure porosity directly in the wellbore, we must deduce it from other measurements.

Accurate knowledge of porosity is important, so hundreds of thousands of research dollars have been spent on developing tools and techniques for measuring *apparent porosity*. (Apparent porosity is the porosity that a particular tool reads in a given formation.) Each of these tools determines a different apparent porosity for the same formation.

The question naturally arises, "Since the different tools give different values for porosity, which one of these values is the correct porosity?" The answer is that all of the tools can accurately determine the porosity of the reservoir rocks—under the proper conditions.

CORES

One common type of porosity measurement is core porosity. To determine core porosity, engineers cut a sample of formation, called a core. A special drilling assembly called a *core barrel* is lowered into the wellbore (Fig. 6–1a). The tool's doughnut-shaped bit drills a hole that leaves a solid plug of formation inside the tool's hollow center. After an appropriate distance is drilled, the tool is hauled back to the surface, retrieving the piece of cored formation inside the assembly. Once the core is released from the barrel, it is packaged and sent to a laboratory where various measurements, including porosity, are made.

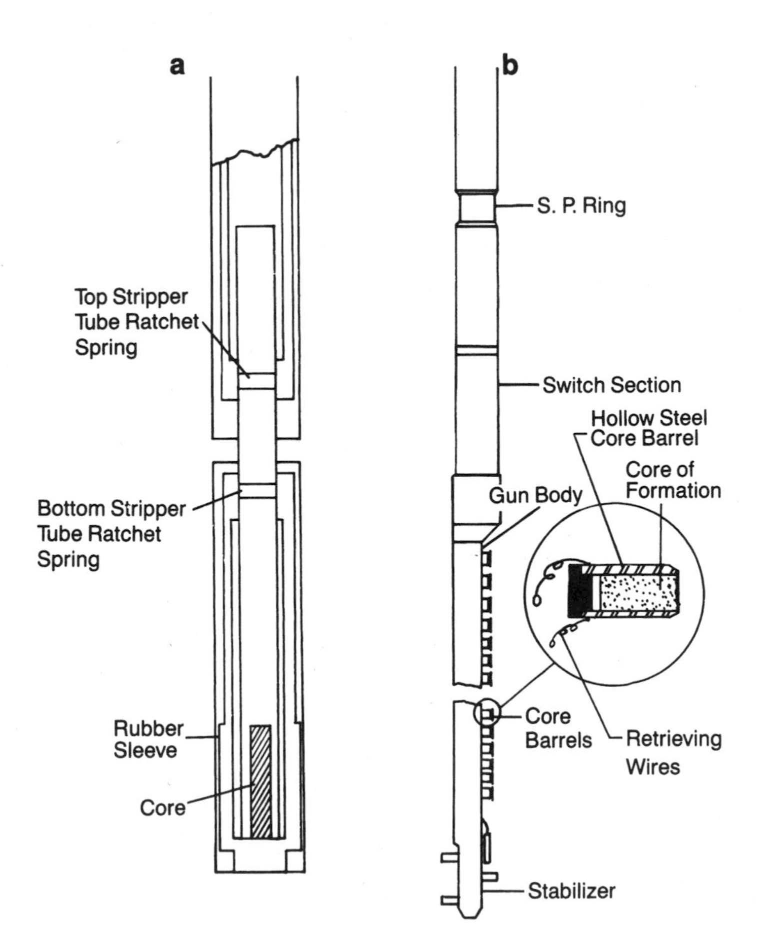

Fig. 6–1. *Coring devices. (a) Core barrel run on drillpipe; (b) sidewall core gun.*

Another method of recovering formation samples is to lower a sidewall core gun into the wellbore on a wireline (Fig. 6–1b). These guns shoot hollow steel bullets into the formation. When the bullets are retrieved, they contain samples of the formation. The core samples are then analyzed in a laboratory for porosity, lithology, permeability, and unusual minerals.

Core porosities may differ from the true formation porosity for several reasons: the rock's properties may be altered during the recovery process; the portion of the core that is measured may not be representative (only small plugs at intervals are actually analyzed); or the volume of rock analyzed may be so small that variations in the formation are missed. Nevertheless, cores are the only way a person can see the formation. With logs, we must use our imagination; with a core, we can actually hold part of the underground rock in our hand.

SUBATOMIC INTERACTIONS

Today it is common to run at least two porosity devices, especially in areas with a mixed lithology of sands, limestones, dolomites, and shales. The two most popular porosity tools are the *compensated density log* and the *compensated neutron log*. Both devices use the formation's response to different types of radioactive bombardment to measure either a density porosity or a neutron porosity. To understand the measurement principles on which these two tools operate, let's digress a bit for a quick lesson in nuclear physics.

The world of nuclear physics deals with matter on a scale unlike anything in common experience. The nuclear physicist works with atoms, neutrons, electrons, protons, positrons, and other subatomic units. All of these units are particles, that is, they have mass.

For the types of interactions that we'll be discussing, we need to be familiar with the following particles:

- **electron**—has a negative charge and a very low mass
- **positron**—just like an electron except it has a positive charge
- **neutron**—electrically neutral but has a mass almost 2000 times that of a positron or an electron
- **proton**—has a positive charge and essentially the same mass as a neutron
- **ray**—nuclear particle that transports energy

- **alpha ray**—helium atom without electrons (2 neutrons + 2 protons)
- **beta ray**—an electron or a positron
- **gamma ray**—a massless particle that travels at the speed of light; also called a photon

Neutrons and protons are usually found in the nucleus of an atom (Fig. 6–2). The number of protons determines the atomic number (*Z*) of an element. The total number of neutrons, protons, and electrons determines the atomic weight (*A*). The number of protons and electrons is generally equal, which results in an electrically neutral atom. (Electrons have a low mass, so they contribute little to the weight or mass of an atom.)

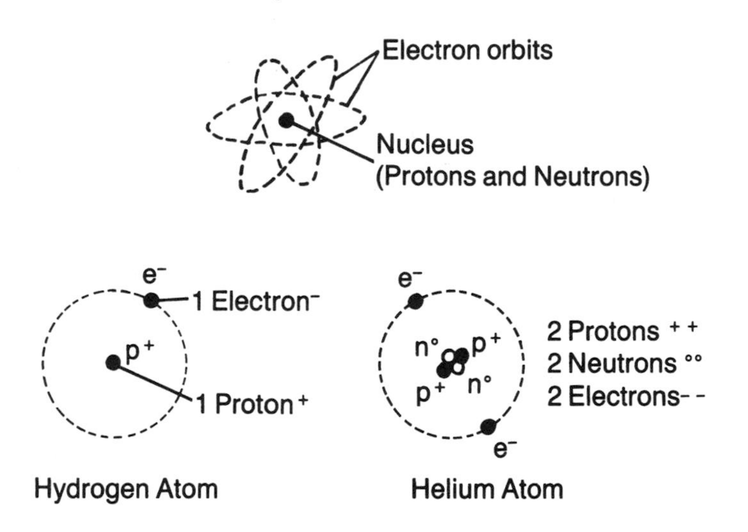

Fig. 6–2. *Structure of the nucleus. The principal subatomic components are electrons, neutrons, and protons.*

If an atom is bombarded by one or more of these particles (neutrons, electrons, alpha rays, etc.), various things can happen, depending on the energy of the bombarding particle, the type of particle, and the amount of energy given up to the atom. When an atom is bombarded by gamma rays, three types of interactions are possible: photoelectric effect, Compton scattering, or pair production (Fig. 6–3).

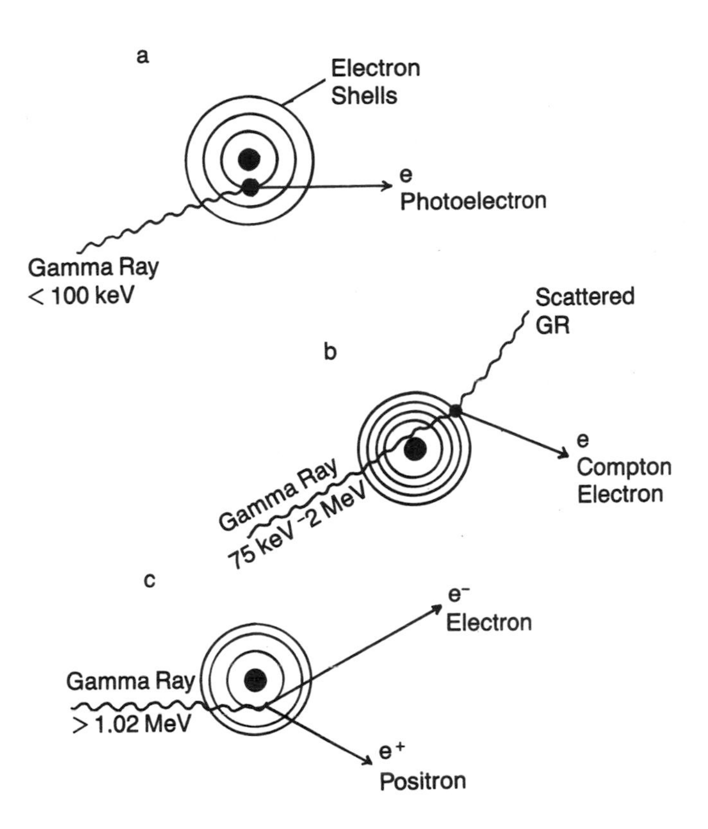

Fig. 6–3. *Gamma-ray interactions. (a) Photoelectric absorption; (b) Compton scattering; (c) pair production.*

In the *photoelectric effect* the gamma-ray energies are < 100 keV (thousand electron-volts). [An electron-volt is a measure of energy. If a particle has an energy of 1000 eV (1 keV), its energy is 1000 times as great as a particle with an energy of 1 eV.] A low-energy gamma ray passes close to the nucleus of an atom and is absorbed completely; an electron is then ejected into space. This reaction is related to the atomic number of the atom and the energy of the *incident,* or impinging, gamma ray. If we know the gamma ray's energy, we can make a measurement proportional to the photoelectric effect. In other words, we can approximate the atomic number. Since we are dealing with a fairly limited number of atoms (primarily silicon, oxygen, calcium, hydrogen, and iron) and compounds in the reservoir rocks, we can calculate the photoelectric effect for different formations. This measurement, then, is indicative of lithology and is largely unaffected by porosity.

If an incident gamma ray has an energy between 75 keV and 2 MeV (million electron-volts), the interactions between the gamma ray and the nuclei are due primarily to *Compton scattering.* In a Compton scattering interaction (an elastic reaction in which both energy and momentum are conserved), a gamma ray hits an electron and imparts some of its energy to that electron. (An elastic reaction is similar to the reactions between billiard balls. When the cue ball strikes another ball, the stationary ball receives some of its energy. If friction forces are disregarded, the total amount of momentum will be the same before and after the collision of the two balls. In other words, momentum of the cue ball before impact equals momentum of the cue ball after impact plus momentum of the struck ball.) The number of interactions is proportional to the number of electrons present in a unit volume. (Unit volume, you'll recall, is a cube one unit long on each side.)

Neutrons are often used as bombarding particles. They are classified according to the interactions they undergo, in much the same way as gamma rays. These interactions correspond to the following energy levels:

- fast neutron: 100,000–15,000,000 eV
- slow neutron: approximately 1000 eV
- epithermal neutron: approximately 1 eV
- thermal neutron: approximately 1/40 eV

The neutrons used in logging come from a source carried in the logging tool. The source contains a type of radioactive material that naturally emits fast neutrons that have elastic reactions while they are in the higher energy ranges. However, with each reaction the neutron loses some of its energy. As a result, the neutron may pass through all of the stages—slow, epithermal, and finally thermal—before it finally loses enough energy to be captured by an atom.

Capture is the other type of interaction that a neutron may undergo. When the neutron is absorbed into or captured by an atom, the atom becomes highly excited (energized) and releases this energy by emitting a gamma ray. This kind of ray is called a *gamma ray of capture.*

The elastic scattering reaction of the high-energy neutrons is sometimes called the *colliding ball reaction.* Imagine that the speeding neutron collides with a stationary atom. If the mass of the atom is much greater than the mass of the neutron, the neutron will bounce off and lose very little energy, like when a golf ball is dropped onto a sidewalk. The massive sidewalk hardly moves at all, but the lightweight golf ball bounces nearly as high as it was dropped; it loses very little of its energy.

On the other hand, if the neutron should collide with something that has essentially the same mass as the neutron, most of its energy will be transferred to the object it strikes. This is like hitting a cue ball into a billiard ball. If the billiard ball is struck straight on, most of the energy will be imparted to the object ball and the cue ball will stop dead.

A neutron has very nearly the same mass as a hydrogen atom. Therefore, the amount of energy that a neutron loses is proportional to the number of hydrogen atoms present. After a few collisions, the neutron is slowed enough to be absorbed or captured by a nearby nucleus. The nucleus then emits a gamma ray of capture. By counting these captured gamma rays, we make a measurement proportional to the number of hydrogen atoms present. Since most formation rock contains no hydrogen atoms, but oil and water do contain hydrogen, the number of hydrogen atoms can be an indication of the porosity.

GAMMA-RAY LOGS

In addition to making measurements attributable to induced radiation, it is also possible to measure the naturally occurring radiation in the wellbore. While the gamma-ray (GR) log is not a porosity log, it is usually run in conjunction with porosity logs (as well as with the resistivity log). The main importance of the GR measurement is to help determine the amount of shale in the formation. Also, the GR curve and the SP curve normally correlate very well since they both respond to the shale content of a formation. The GR curve is recorded in track 1.

The GRs that we measure with this tool are the naturally occurring gamma rays rather than the induced gamma rays caused by a radioactive logging source, as in the density tool. These natural gamma rays emanate from radioactive potassium, thorium, and uranium, the three elements that account for most of the radiation in sedimentary formations. Potassium and thorium are closely associated with shale (illite, kaolinite, montmorillonite), while uranium may be found in sands, shales, and some carbonates.

On the whole, however, the GR curve is almost unaffected by porosity and is an excellent indicator of shale. By using the relative response of the curve compared to the reading in a 100% shale (GR_{sh}), we can estimate the volume of shale (V_{sh}) in the formation from the equation

$$V_{sh} = GR_{sh} - GR_{zone} / GR_{sh} - GR_{clean}$$

where:

V_{sh} = volume of shale
GR_{sh} = gamma ray value at the shale point
GR_{zone} = gamma ray value in the zone of interest
GR_{clean} = the lowest gamma ray value in a nearby zone

Look ahead to the shale baseline at 8 divisions in Figure 6–4. This is GR_{sh}. It is drawn through the average reading of a thick, uniform shale. (Do not use the maximum readings; these usually correspond to high concentrations of uranium compounds.) Also note the minimum reading at point A (0.8 divisions). This is the 100% clean or shale-free point (GR_{clean}).

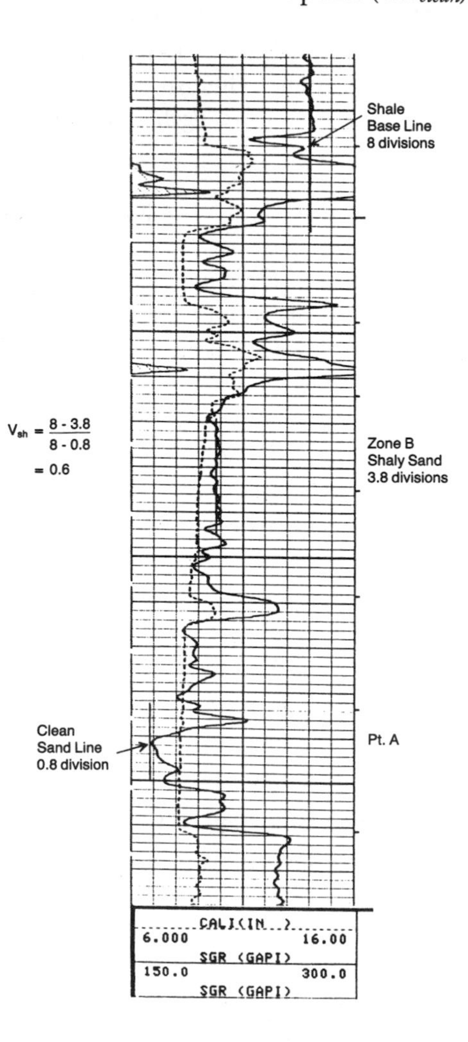

Fig. 6–4. *Gamma-ray log. This example shows how to calculate shale volume, V_{sh}.*

To determine V_{sh} of any zone, such as zone B, subtract the clean sand reading from the shale baseline reading. This is the denominator in the V_{sh} equation. Next, read the log at the zone for which V_{sh} is required (3.8 divisions for zone B). Subtract this reading from the shale baseline reading; this is the numerator. Divide the numerator by the denominator; the result is V_{sh}. In our example, $GR_{sh} = 8$; $GR_{clean} = 0.8$; $GR_{zone} = 3.8$. The equation becomes

$$V_{sh} = (8 - 3.8)/(8 - 0.8) = (4.2)/(7.2)$$
$$= 0.6$$

The original GR tools measured the total natural radiation present in the wellbore. Detector technology was too primitive to separate the different energy levels of the natural radiation into the individual contributions made by radioactive elements present. Today, by using much more sensitive detectors and improved tool designs with multiple windows, the natural gamma rays are divided into the parts contributed by each element. This division into parts is based on the assumption that all of the radiation is from thorium, potassium, or uranium—the three elements that account for most of the natural radiation present in the formations we are likely to encounter when drilling a well. By measuring these three elements, we are able to make a more accurate estimate of shale content. Most of the uranium is found in sandstones. By removing the uranium contribution from the total GR curve, we can calculate the percent of shale more accurately.

DENSITY LOGS

Density is the weight of a unit volume of a substance. For example, 1 ft.^{3} of distilled water weighs 62.4 lb and thus has a density of 62.4 lb/ft.^{3} in the English system of measurement, while pure limestone's density is 169 lb/ft.^{3}. In the metric measuring system, 1 cc of water weighs 1 g. So water has a density of 1.0 g/cc and limestone's density is 2.71 g/cc. For logging in the United States, we measure in grams per cubic centimeter (g/cc).

Unfortunately, we can't measure formation, or bulk, density directly from the borehole. However, we can measure electron density by using Compton scattering reactions, and electron density is very nearly the same as bulk density.

The density-measuring tool (Fig. 6–5) bombards the formation adjacent to the wellbore with gamma rays from a cesium source and Compton scattering takes place. The gamma rays are counted by two detectors mounted on a skid pressed against the borehole wall. The two detectors compensate for the effects of hole rugosity and mud-cake thickness while the bulk density measurement is made. (That's why the tools are called compensated density tools.)

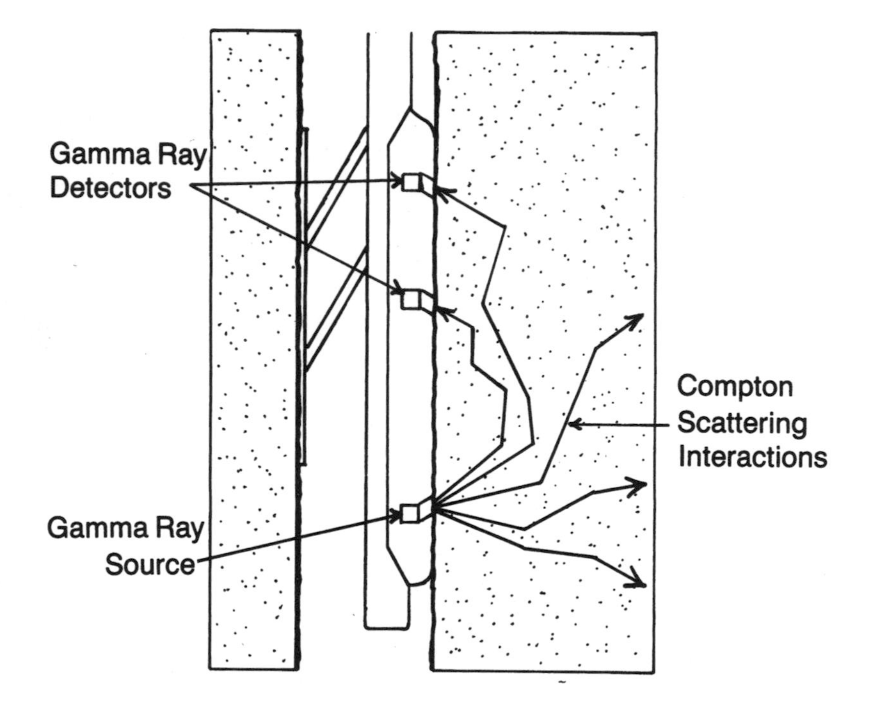

Fig. 6–5. *Compensated formation density tool. Gamma rays emitted by the source undergo Compton scattering reactions within the formation. The two detectors compensate for hole rugosity.*

The newest generation of density tools measures the photoelectric absorption cross-section of the formations. The photoelectric effect is another way the formation reacts to the bombarding gamma rays. This reaction occurs at a much lower energy level than Compton scattering. By measuring the energy level of the formation reactions, engineers can separate photoelectric reactions from other reactions. The photoelectric response is then used to help identify lithology.

Interpretation—Density Log

The density (ρ) determined by the density tool is called RHOB (ρ_b), where *b* stands for bulk volume. The rock structure of minerals such as sandstone or limestone is called the matrix. The density of this structure is called the matrix density, ρ_{ma}. This is the density that the tool would read if the formation had zero porosity. The density of the fluid in the pore space, usually mud filtrate, is called ρ_f.

A very important principle of mathematics is that the whole is equal to the sum of its parts. We use this axiom constantly in log interpretation; determining porosity from density measurements is a good illustration. The bulk volume of formation is the whole, and the parts are the matrix volume and the fluid volume contained in the pore space. Let's assume that we know the matrix density (the densities of sandstones, limestones, dolomites, and other minerals have been measured in the laboratory) and the fluid density. Then we can write an equation based on the principle that the whole equals the sum of its parts:

density log reading (ρ_b) = (matrix volume (BVM) X matrix density (ρ_{ma}))
+ (fluid volume X fluid density (ρ_f))

We know that the matrix volume is (1– ϕ) and the fluid volume is ϕ, so

$$\rho_b = ((1 - \phi) \times \rho_{ma}) + (\phi \times \rho_f)$$

To apply this principle to the earth's formations in order to determine porosity, we must know or assume (1) the density of the matrix and (2) the density of the fluid in the pore spaces. We can often make these assumptions in developed fields where the lithology is accurately known or in regions that are predominantly sandstone, such as the U.S. Gulf Coast or California.

We know from experience and laboratory measurements that the matrix density is 2.71 g/cc for limestone, 2.87 g/cc for dolomite, and for sandstones 2.65 g/cc (unconsolidated) or 2.68 g/cc (mature). The fluid in the pore space is water, oil, or gas. Since the density tool has a shallow depth of investigation, it is likely that the formation will be filled mainly with mud filtrate. The fluid density is therefore generally assumed to be 1.0 g/cc, but corrections can be applied to the value if necessary.

Figure 6–6 is a section of a Litho-Density log (a mark of Schlumberger). Note the scales and curves in tracks 2 and 3. On the log, the density curve (solid) is designated RHOB (ρ_b) and is scaled from 2.0 to 3.0 g/cc. The dashed curve labeled PEF and scaled 0.0 to 10.0 is the photoelectric index curve. The dotted curve in track 3 is the density correction curve (DRHO for $\Delta\rho$). It monitors the amount of correction being added to the ρ_b curve by the compensation circuitry. If $\Delta\rho > 0.15$, use the density reading with caution because the correction is excessive and the density reading may be incorrect. The heavy dashed line between divisions 7 and 8 of track 3 is the tension curve, which monitors the drag when the logging tool rubs against the side of the hole.

To use the density curve, we need a chart that converts ρ_b to porosity (Fig. 6–7). To use the chart we need to know the lithology. That's where the PEF curve comes in. Since the PEF for sandstone is 1.8, limestone is 5.1, and dolomite is 3.1 (Table 6–1), we can use the PEF curve to identify the matrix minerals. The only obstacle is shale, whose PEF can range from 1.8–6.3 but is usually around 3, like that of dolomite.

Shale has a slight effect on the density porosity readings, so its influence must be removed. To remove the shale effect, follow these steps:

1. Find a uniform shale section.
2. Read the apparent density porosity in the shale.
3. Apply the shale correction equation:

$$\Phi_{Dcor} = \Phi_D - (V_{sh} \times \Phi_{Dsh})$$

where:

Φ_{Dcor} = corrected density porosity
Φ_D = density porosity from the log
V_{sh} = shale volume
Φ_{Dsh} = apparent density porosity in the shale

In practice, the effect of shale on the density log is often ignored unless V_{sh} is high (> 30%).

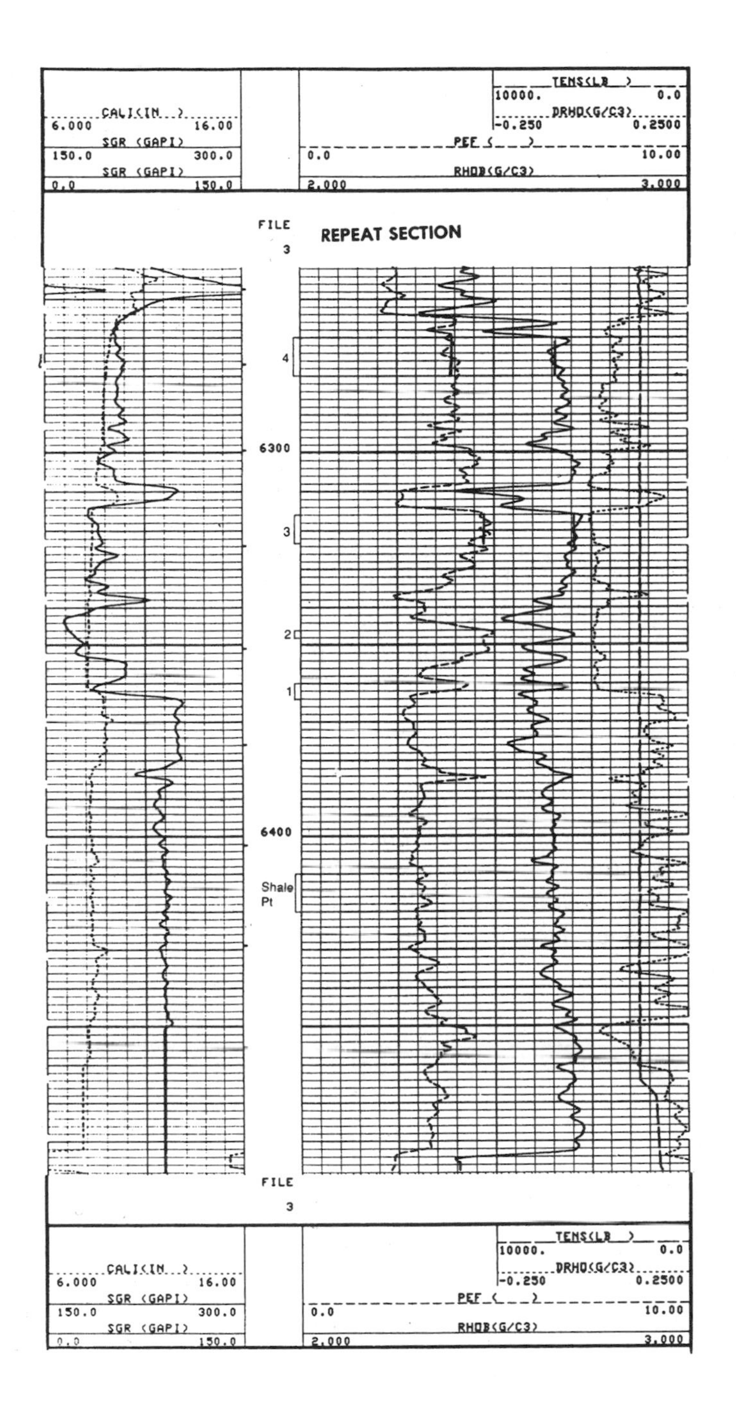

Fig. 6–6. *Section of a Litho-Density log (courtesy Schlumberger).*

Now it's time for some practice in converting density readings to porosities. Set up a table so you can record the values from the log in Figure 6–6. The table should look like this:

Point No.	PEF	ρ	Δρ	Porosity	Lithology
1	4.3	2.69	+0.01		
2					
3					
4					

Now read the values (except porosity; you'll find that from the chart) from the log at these points. You're already started with point 1. If you have trouble reading the log, review Chapter 2. If you just want to check your answers, see the end of this chapter for our picks, complete with porosity and lithology estimate.

Once you've made your log picks, turn to Figure 6–7. Enter point 1 at the bottom of the chart at PEF = 4.3. Draw a vertical line. Next enter ρ_b = 2.69 at the left edge and draw a horizontal line until it intersects the PEF line. This line plots between the dolomite line and the limestone line. To determine the porosity, connect points of equal porosity on the two matrix lines (10 to 10, 5 to 5, etc.). Point 1 falls on the 5% line and is closer to the limestone line than to the dolomite line, so enter 5% and dolomitic limestone in the table.

Point No.	PEF	ρ	Δρ	Porosity	Lithology
1	4.3	2.69	+0.01	5%	Dolomitic limestone
2					
3					
4					

Try your hand at the other three points; then check your answers. Although adding the PEF curve to the density tool helps determine both porosity and lithology under favorable conditions, sometimes the PEF curve isn't available (slim tools), doesn't work (heavy mud weights), or wasn't run (older density tools). In those cases, we either assume the lithology or we require additional information that can only be provided by another porosity measurement. One of these tools is the compensated neutron log.

COMPENSATED NEUTRON LOG

The original neutron tool was an early development. By bombarding the formation with neutrons from a chemical source in the logging tool, engineers could measure the response of the formation as a function of the number of hydrogen atoms present. Because most of the hydrogen present is in the water (H_2O) and oil (C_2H_{2n+2}) and because one or both of these fluids are present in the pores of the rocks, we can determine the porosity simply by counting the hydrogen atoms.

This is what the single-detector neutron tool does. However, it indicates rather than measures porosity. The response of the tool is nonlinear; it has high resolution in low porosities but very little resolution in high porosities. For this reason, it is used mainly in hard-rock areas (low porosity) and as a correlation tool in casing. (The steel in the casing prevents most tools from making a valid measurement, but both the natural gamma-ray and the neutron tool will read through the casing. Often these two measurements are used to correlate the open-hole measurements to some feature of the casing, such as the depth of the casing collars. The casing collars, which are easily detected, may then be used as a reference for positioning perforating guns, plugs, packers, etc.)

In the early 1970s, a major development was made in neutron tool instrumentation, resulting in the compensated neutron (CN) device. The CN tool uses two detectors to compensate for hole rugosity. In addition, it measures the ratio of the detector responses and converts this ratio to a linear porosity reading instead of the nonlinear response of the single-detector neutron tool.

Interpretation—Compensated Neutron Log

The compensated neutron log is generally recorded on a limestone matrix. Thus, the porosity readings shown on the log are correct if

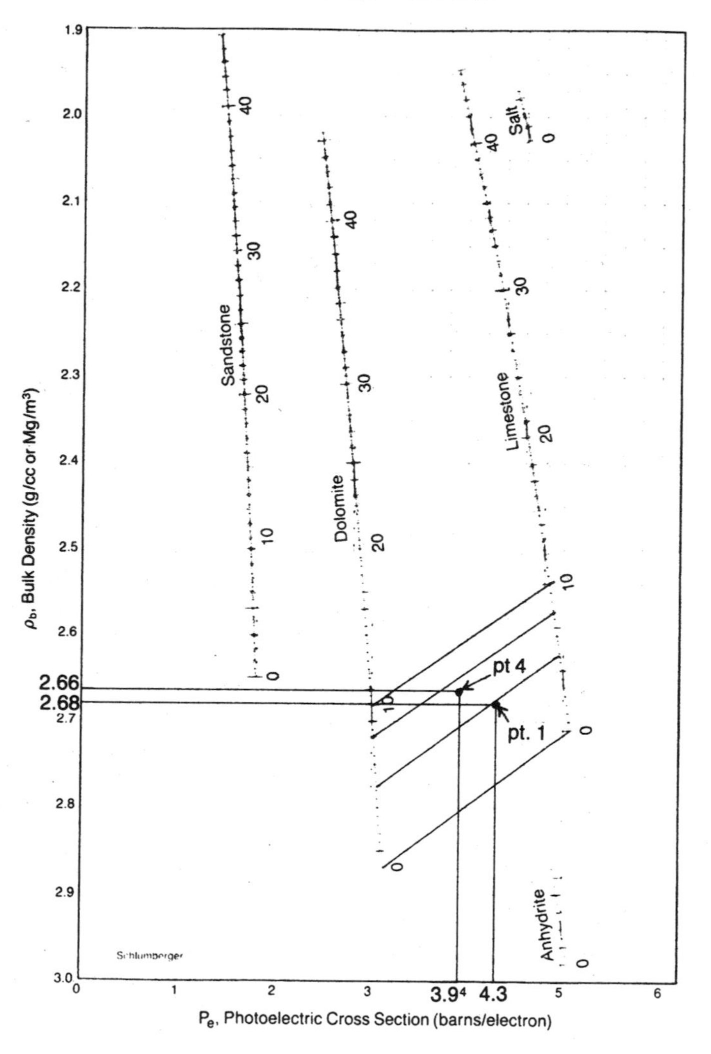

Fig. 6–7. *Chart to determine porosity and lithology from a Litho-Density log (courtesy Schlumberger).*

- the formation is clean (no shale)
- the porosity is filled with liquid (not with gas)
- the formation is limestone

Remember that the neutron tool responds principally to the number of hydrogen atoms present. Also remember that shale has a large number of hydrogen atoms because water molecules are bound to the clay, but its effective porosity is essentially zero due to the very fine grain size. The presence of shale in a formation increases the total porosity, but the effective porosity remains the same. Since we're interested in effective porosity and not total porosity, we must subtract the shale value from the total porosity reading. We do this by applying the following equation:

$$\Phi_{Ncor} = \Phi_N - (V_{sh} \times \Phi_{Nsh})$$

where:

Φ_{Ncor} = corrected neutron porosity
Φ_N = apparent neutron porosity read from the neutron log
V_{sh} = shale volume
Φ_{Nsh} = apparent neutron porosity at the shale point

The neutron porosity should be corrected for shale whenever V_{sh} exceeds 5%.

The presence of gas in the formation also has a pronounced effect on the neutron log readings. Since gas has many fewer hydrogen atoms per unit volume than either water or oil, the apparent porosity in a gas zone is much lower than it should be. To correct for the gas effect, we need to know density and/or sonic porosity.

If the formation is not limestone but is sandstone or dolomite, we must correct the apparent limestone porosity to the proper matrix using Figure 6–8. However, we can use this chart only if we know the matrix. If we don't know the matrix, we need another porosity measurement to determine both porosity and lithology.

Another type of neutron porosity tool is the SNP or sidewall neutron porosity device, so called because the detector and the source are mounted on a skid similar to the density skid and are pressed against the side of the borehole. One of the advantages of the SNP is that it can be run in empty or air-filled holes. The CN log does not respond properly in those conditions, but the SNP does. Before the development of the CNL, the SNP was often used in mud-filled holes. However, today the SNP is generally used only in air-drilled holes.

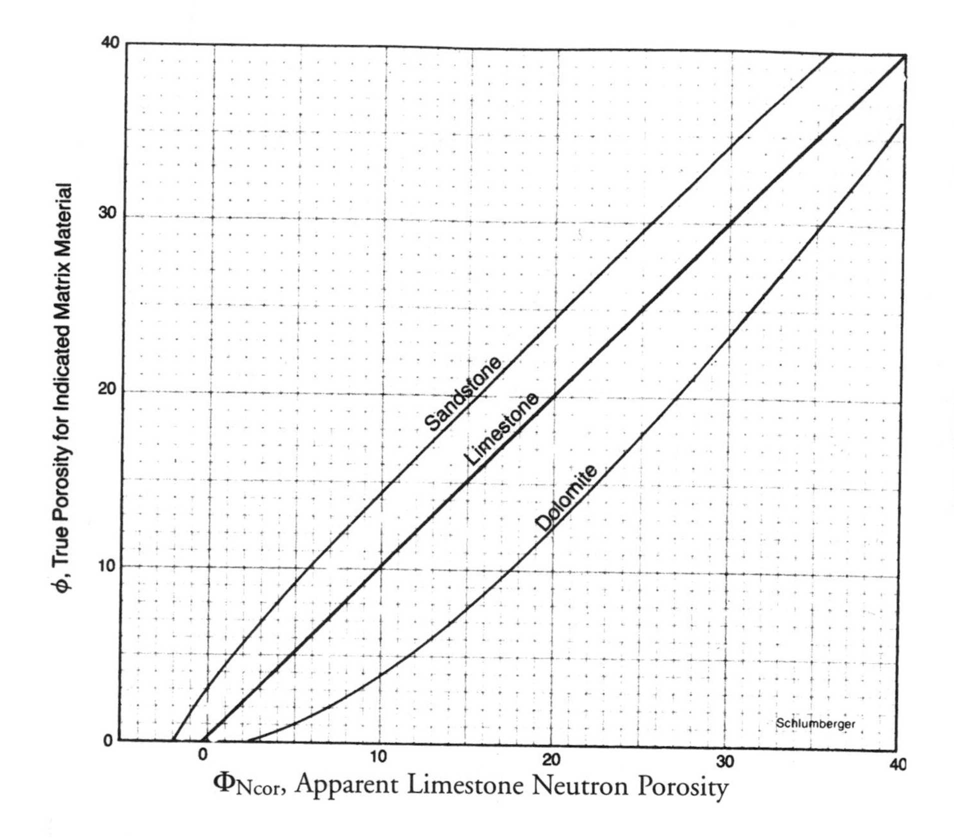

Fig. 6–8. *Neutron porosity equivalence curves (courtesy Schlumberger). Usually the neutron log is recorded in limestone porosity units (pu's). Use the chart to convert to other rock types.*

SONIC LOG

The *sonic* or *acoustic tool* uses sound waves to measure porosity. Let's take a quick look at sound waves and how they travel.

Sound is energy that travels in the form of a wave and has a frequency between 20 and 20,000 cycles per second (cps, or hertz). A sound wave (also called an acoustic wave) can travel in several different forms. The most common form is a compressional wave, the kind of wave that vibrates our eardrums so we can hear. Compressional waves are also called *P*-waves (primary waves) because they are the first waves to arrive.

A *P*-wave travels by compressing the material in which it travels. The material "moves" along the axis of the wave. An example of a *P*-wave is a Slinky™ spring toy that you hold outstretched vertically. If you lift a couple of coils and then drop them, a compressional wave will travel down the spring. When the wave reaches the end of the coil, it will travel back up. This phenomenon is called *reflection.* Sound waves also change speed when the material in which they are traveling changes, a process called *refraction.*

A second type of sound wave is the shear wave, or *S*-wave. This wave is slower than the *P*-wave and cannot be transmitted through a fluid. To visualize an *S*-wave, think of a rope with one end tied to a tree. If you pull the free end of the rope almost tight and then snap it, a shear wave will roll down the rope. The rope does not move horizontally; it moves vertically, or at right angles to the axis of the wave. This motion is characteristic of an *S*-wave. If it were not in a solid medium, it would not be able to transmit its energy.

Several other types of waves may be present in a full-wave recording of sound passing through the formation near the borehole. Some of these waves, such as Stoneley waves, are used to determine permeability and fracturing. Research continues with variable-frequency sound sources and increased data sampling, making acoustic energy tools more useful than ever.

The sonic tool takes advantage of the fact that a sound wave travels at different speeds through different materials and, more importantly, that the sound wave travels at different speeds through mixtures of materials. If we know the speed of sound for each of the materials, we can calculate the amount of each material as long as there are only two materials. If there are more than two substances, we need additional information. In other words, if we know that a certain formation is a limestone and that any pore spaces it may have are filled with water, we can determine the porosity by measuring the time a compressional sound wave takes to travel through 1 ft. of the formation.

Figure 6–9 is a schematic of a sonic tool that has one transmitter and two receivers. The transmitter is 3 ft. from the first receiver and 5 ft. from the second receiver, and it emits a strong sound pulse that travels spherically outward in all directions. The mud column and the tool have slower traveltimes (sonic velocities) than the formations.

The first sound energy to arrive at the two receivers is the *P*-wave, which travels through the formation near the borehole. The difference in the times at which the signal reaches the two receivers is divided by the spacing of the receivers. This time, recorded in microseconds per foot, is also called sonic interval traveltime (t) for the difference in arrival times between the two receivers.

In practice, the tools that measure t are much more complicated than the tool illustrated here. They have multiple transmitters and receivers to compensate for sonde tilt, washed-out hole, and alteration of the rock properties near the wellbore due to drilling processes.

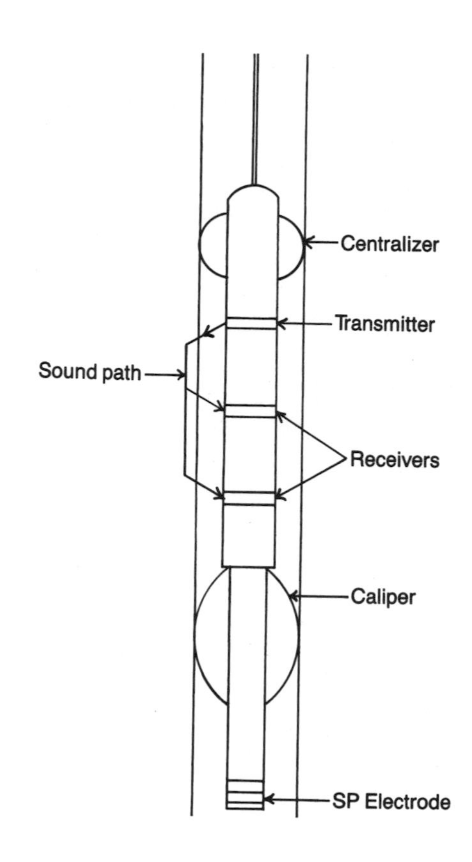

Fig. 6–9. *Simplified schematic of a sonic tool. The difference in arrival time, Δt, between the receivers measures the formation travel time.*

Interpretation—Sonic Log

If we know the interval traveltime and the type of formation and can assume the porosity is uniformly distributed (intergranular as opposed to vugular or fracture porosity), we can determine the porosity (Fig. 6–10). As with any single porosity measurement, we must know or assume the lithology to make an estimate.

Shale has a strong effect on the sonic log. In shaly formations ($V_{sh} > 5\%$) the sonic porosity must be corrected for the presence of shale with the equation

$$\Phi_{Scor} = \Phi_S - (V_{sh} \times \Phi_{Ssh})$$

where:

Φ_{Scor} = corrected sonic porosity
Φ_S = sonic porosity determined from Figure 6–9
V_{sh} = shale volume
Φ_{Ssh} = apparent porosity of the shale point

Gas also has a strong effect on the apparent sonic porosity; it raises the apparent porosity. If gas-bearing formations are anticipated, engineers should run at least one other porosity device, preferably a compensated neutron log.

Unconsolidated sandstones such as those in California or on the Gulf Coast have longer traveltimes than they should for their porosity. To correct the traveltimes for unconsolidation, the traveltime reading (t_{sh}) at the shale point is used to determine the compaction correction (B_{cp}):

$$B_{cp} = t_{sh}/100$$

For example, if t_{sh} = 120 µs/ft., then B_{cp} is 120/100 = 1.2.

Instead of the sandstone line (55 µs/ft.) in Figure 6–10, the correct B_{cp} line is used for entering traveltime from the log whenever the formation is unconsolidated.

Finally, the sonic wave does not detect the porosity in vugs and fractures (secondary porosity) as well as it detects intergranular (primary) porosity. This lowers the apparent porosity in vugular and/or fractured formations. By comparing the sonic primary porosity to the total porosity from other logs, we can estimate the amount of fracture or vugular (secondary) porosity.

MULTIPLE POROSITY LOGS

We have already seen how to derive porosity from individual porosity logs. With a single device, we must assume we know the mineralogy of the formation; unfortunately, this is seldom the case. Both gas and shale make interpreting single porosity measurements less reliable. However, a powerful technique exists that helps us overcome many of the limitations of the single porosity measurement: the *porosity crossplot technique.*

With the crossplot technique and two porosity measurements, we can determine porosity that is independent of lithology, i.e., we don't have to assume the lithology. Since three different porosity tools are available (density, neutron, and sonic), we can combine crossplots. These plots, called mineral identification plots, allow a very accurate estimate of rock type. The most common combination of measurements is made with the compensated neutron density log.

When the density and the neutron logs are run together, they are usually recorded on a limestone matrix as though all the formations were limestones. (It is very important to check the matrix type used to record the logs; the matrix chosen varies from region to region.) In a shale-free, wet (S_w = 100%) limestone, the two logs will read the same porosity. If the zone is a

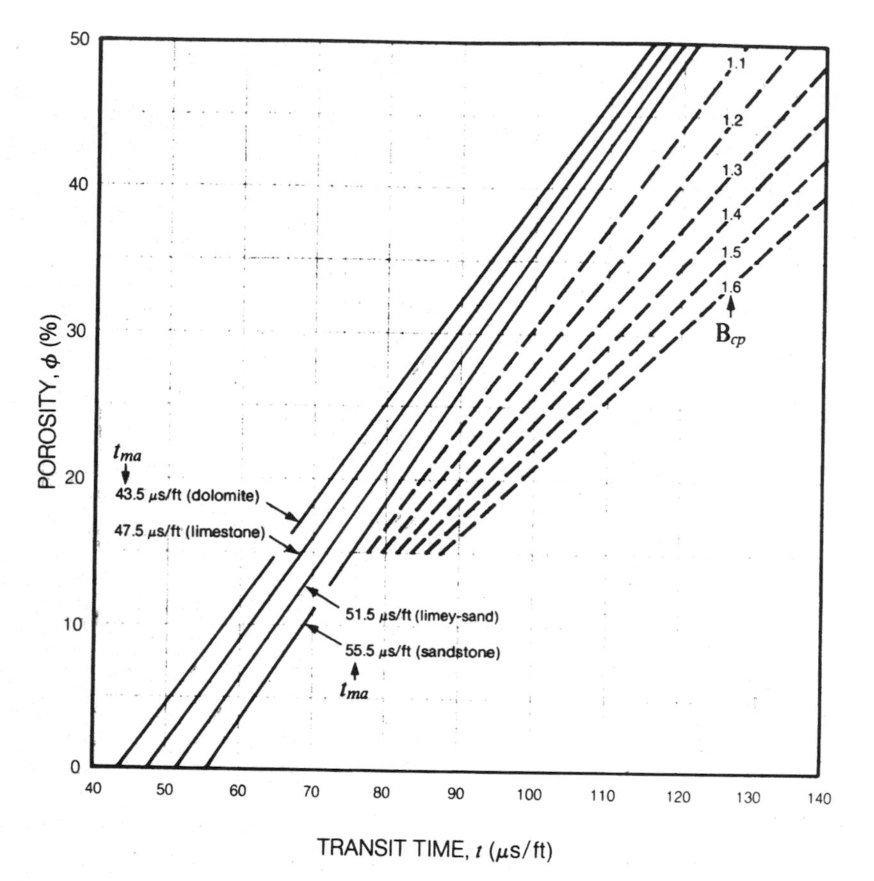

Fig. 6–10. *Determining porosity from sonic interval traveltime (courtesy Schlumberger). Sonic interval traveltimes (t) may be converted to porosity if the lithology is known.*

wet sandstone, the neutron porosity will read too low and the density porosity will read too high. There will be a positive separation between the two curves, which is always of interest. The separation could be due to a different matrix or to the presence of gas. Gas is usually indicated if the separation is > 6 pu. To verify the presence of gas, a third porosity measurement is necessary (or the lithology must be assumed). Shale and dolomite also cause the density and neutron porosity curves to separate, but in the opposite direction from gas or sandstone. In a shale or dolomite, the neutron porosity will be higher than the density porosity. A third porosity device, such as the sonic, is usually necessary to determine whether the separation is from shale or dolomite.

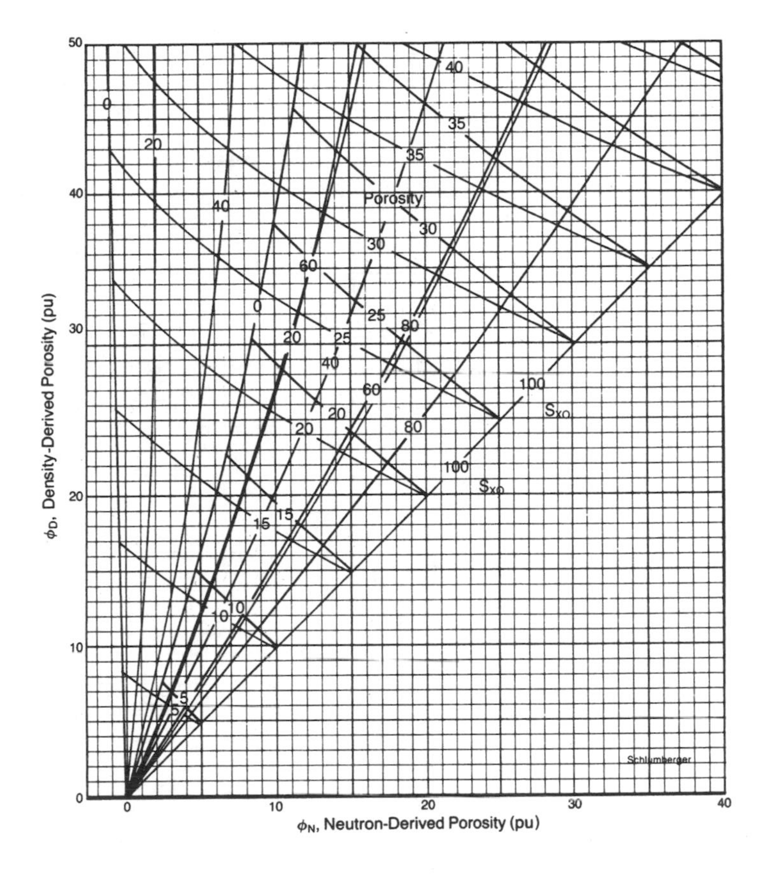

Fig. 6–11. *Gas-bearing formation porosity from density and neutron logs (courtesy Schlumberger).*

With three porosity logs, lithology as well as porosity can be determined accurately. Once lithology is known, the density and the neutron values can be corrected by using the proper matrix and entered in Figure 6–11. This chart yields the crossplot porosity free of matrix effects and an estimate of S_{xo}.

Density and neutron tools are often *stacked* (connected so they can be run simultaneously) when logging wells with unknown or mixed lithology or in gas wells. Figure 6–12 is the neutron density log over the same interval of formation that we examined with the Litho-Density tool in Figure 6–6. By entering the neutron (Φ_N) and density (Φ_D) porosities in Figure 6–13, we can estimate the lithology and plot the porosity.

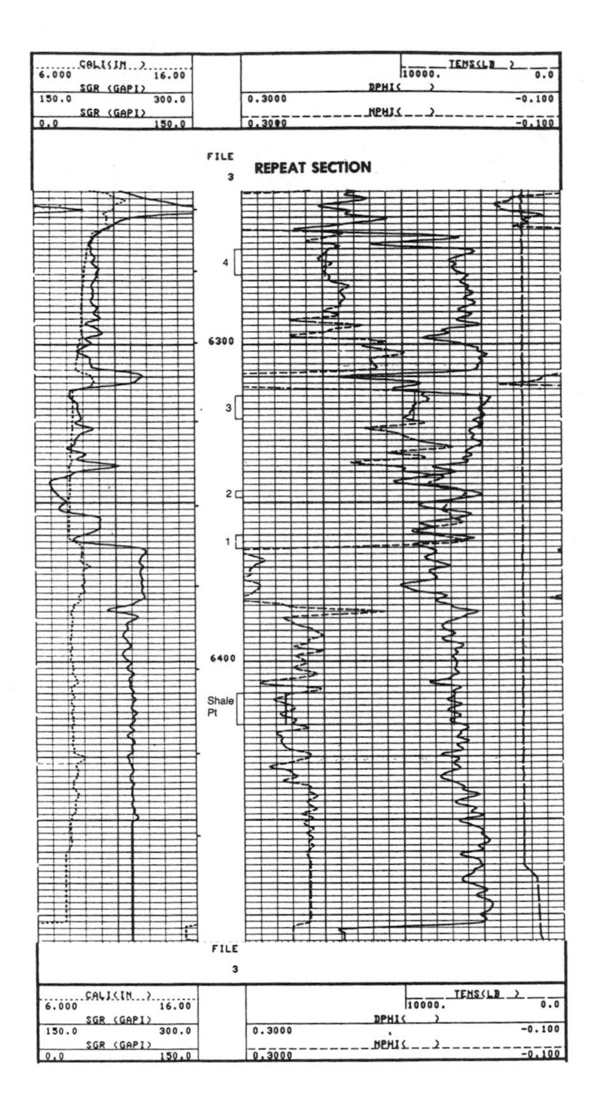

Fig. 6–12. *Sample compensated neutron density log.*

Let's try the same points as before. This time, though, we must correct for shale on both the density and neutron readings:

$$\Phi_{Ncor} = \Phi_N - (V_{sh} \times \Phi_{Nsh})$$
$$\Phi_{Dcor} = \Phi_D - (V_{sh} \times \Phi_{Dsh})$$

The density porosity (Φ_D) is the solid curve and the neutron porosity (Φ_N) is the dashed curve. Both curves are recorded on a limestone matrix and are scaled 30% to –10%. First let's set up a table.

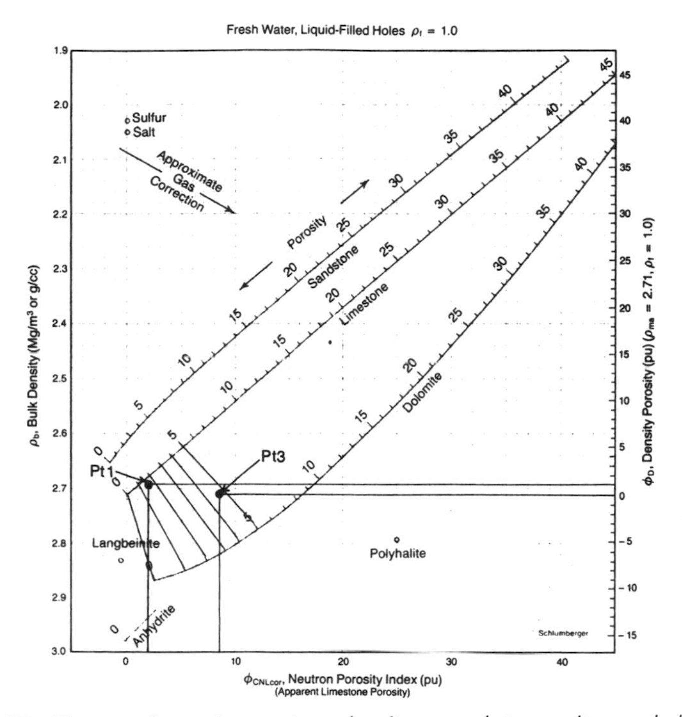

Fig. 6–13. *Chart to determine neutron density crossplot porosity, used along with Figure 6–12.*

Point No.	V_{sh}, %	Φ_D, %	Φ_N, %	Φ_{Dcor}, %	Φ_{Ncor}, %	Φ_{xp}, %	Lithology
Shale	100	3.5	25	—	—	—	Shale
1	5	1	2	1	1	1	Limestone
2	0						
3	25						
4	50						

Now that you're started, fill out the second and third columns; then find the crossplot porosity (Φ_{xp}). (Answers are given at the end of the chapter.) Note that there is good agreement between the Litho-Density tool and the neutron density crossplot, except for point 4, because the PEFs for dolomite and shale are often similar.

To find the crossplot porosity, follow these steps:

1. Make the shale corrections to Φ_N and Φ_D.
2. Enter the corrected density and neutron porosities Φ_{Ncor} and Φ_{Dcor} in the proper columns of the data sheet.

3. On the right-hand side of Figure 6–13, find Φ_{Dcor} and draw a horizontal line on the chart (at 1% density porosity for point 1).
4. Draw a vertical line through Φ_{Ncor} (1% in this case).
5. Read the crossplot porosity and lithology at the intersection of the two lines. (Point 1 falls almost exactly on the limestone matrix line at 1% porosity.)

If the intersection does not fall very close to one of the matrix lines, assume the formation is a mixture of the two matrices that the point lies between (see point 3). To help estimate the porosity, connect equal porosity points as in Figure 6–13.

QUICK-AND-DIRTY CROSSPLOT POROSITY

The method just explained is the most accurate manual way of determining the crossplot porosity. However, there is a faster, although somewhat less accurate, way that does not use charts.

1. Read the porosity from the logs.
2. Add the two porosities and divide by 2.
3. Correct the averaged porosity for shale:

$$\Phi_e = \Phi_{avg} - (V_{sh} \times \Phi_{shxp})$$

where:

Φ_e = effective porosity
Φ_{avg} = averaged porosity
V_{sh} = shale volume
Φ_{shxp} = averaged apparent porosity of shale point

Let's see how the faster method compares with the chart method. Set up a table again and calculate the crossplot porosities for points 1 through 4. You already have a head start.

Point	V_{sh}, %	Φ_D, %	Φ_N, %	Φ_{avg}, %	$V_{sh} \times \Phi_{sh}$, %	Φ_{xp}, %
Shale	100	3.5	25	14	14	0
1	5	1	2	1.5	0.7	1*
2	0					
3	25					
4	50					

** Round off porosity to the nearest half porosity unit such as 5.5 or 14, not 5.36 or 14.1.*

Compare the three different porosity tables at the end of the chapter. There is little difference; but note that with the faster quick-and-dirty method you do not get a lithology estimate.

As a final review to this chapter, note Table 6–1, which lists various minerals and some of their log-derived physical parameters such as density, photoelectric index, neutron porosity, sonic compressional and shear traveltimes, capture cross-section, dielectric constant, and natural gamma ray. Then, after you check your answers to the problems, turn to Chapter 7 so you can put all your newfound knowledge to practical use.

Answers to Problems

Density only:

Point	Pef	ρ_b RHOB	$\Delta\rho$ DHRO	Porosity, %	Lithology
1	4.3	2.60	+0.01	5	Dolomitic limestone
2	4.9	2.71	+0.005	0.5	Limestone
3	4.7	2.71	0.00	1	Limestone
4	3.9	2.66	+0.07	8	50/50 limestone-dolomite

Neutron-density crossplot:

Point	V_{sh}, %	Φ_D, %	Φ_N, %	Φ_{Dcor}, %	Φ_{Ncor}, %	Φ_{xp}, %	Lithology
Shale	100	3.5	25	–	–	–	Shale
1	5	1	2	1	1	1	Limestone
2	0	0	2	0	2	1	Limestone
3	25	0	8.5	0	2	1	Limestone
4	50	1.3	19.5	1.5	7	4.5	Limestone-dolomite

"Quick-and-dirty":

Point	V_{sh}, %	Φ_D, %	Φ_N, %	Φ_{avg}, %	V_{sh} x Φ_{sh}, %	Φ_{xp}, %
Shale	100	3.5	25	14	14	0
1	5	1	2	1.5	0.7	1
2	0	0	2	1	0	1.5
3	25	0	8.5	4.2	3.5	1
4	50	3	19.5	11	7	4

NAME	FORMULA	ρ_b, g/cc	Φ_{SNP}, pu	Φ_{CNL}, pu
CLAYS				
Kaolinite	$Al_4Si_4O_{10}(OH)_8$	2.41		37
Chlorite	$(Mg,Fe,Al)_6(Si,Al)_4O_{10}(OH)_8$	2.76	37	52
Illite	$K_{1-1.5}\ Al_4(Si_{7-6.5},Al_{1-1.5})O_{20}(OH)_4$	2.52	20	30
Montmorillonite	$(Ca,Na)_7(AI,Mg,Fe)_4$ $(Si,Al)_8O_{20}(OH)_4(H_2O)n$	2.12	40	44
EVAPORITES				
Halite	$NaCl$	2.04	–2	–3
Anhydrite	$CaSO_4$	2.98	–1	–2
Gypsum	$CaSO_4(H_2O)_2$	2.35	50+	60
Trona	$Na_2CO_3NaHCO_3H_2O$	2.08	24	35
Tachydrite	$CaCl_2(MgCl_2)_2(H_2O)_{12}$	1.66	50+	60
Sylvite	KCl	1.86	–2	–3
Carnalite	$KClMgCl_2(H_2O)_6$	1.57	41	60
Lanqbenite	$K_2SO_4(MgSO_4)_2$	2.82	–1	–2
Polyhalite	$K_2SO_4MgSO_4(CaSO_4)_2(H_2O)_2$	2.79	14	25
Kainite	$MgSO_4KCl(H_2O)_3$	2.12	40	60
Kieserite	$MgSO_4H_2O$	2.59	38	43
Epsomite	$MgSO_4(H_2O)_7$	1.71	50+	60
Bischofite	$MgCl_2(H_2O)_6$	1.54	50+	60
Barite	$BaSO_4$	4.09	–1	–2
Celesite	$SrSO_4$	3.79	–1	– 1
SULFIDES				
Pyrite	FeS_2	4.99	–2	–3
Marcasite	FeS_2	4.87	–2	–3
Pyrrhotite	Fe_7S	4.53	–2	–3
Sphalerite	ZnS	3.85	–3	–3
Chalcopyrite	$CuFeS_2$	4.07	–2	–3
Galena	PbS	6.39	–3	–3
Sulfur	S	2.02	–2	–3
COALS				
Anthracite	$CH_{.358}N_{.009}O_{.022}$	1.47	37	38
Bituminous	$CH_{.793}N_{.015}O_{.078}$	1.24	50+	60
Lignite	$CH_{.849}N_{.015}O_{.211}$	1.19	47	52

t_c µs/ft.	t_s µs/ft.	P_e barn/elect	U barn/cc	ε farads/m	t_p nsec/m	GR API units	Σ c.u.
		1.83	4.44	~5.8	~8.0	80–130	14.12
		6.30	17.38	~5.8	~8.0	180–250	24.87
		3.45	8.73	~5.8	~8.0	250–300	17.58
		2.04	4.04	~5.8	~8.0	150–200	
67	120	4.65	9.45	5.6–6.3	7.9–8.4	–	754.2
50		5.05	14.93	6.3	8.4	–	12.45
52		3.99	9.37	4.1	6.8	–	18.5
65		0.71	1.48			–	15.92
92		3.84	6.37			–	406.02
		8.5	15.83	4.6–4.8	7.2–7.3	500+	564.57
		4.09	6.42			–220	368.99
		3.56	10.04			–290	24.19
		4.32	12.05			–200	23.70
		3.50	7.42			–245	195.14
		1.83	4.74			–	13.96
		1.15	1.97			–	21.48
100		2.59	3.99			–	323.44
		266.82	1091			–	6.77
		55.19	209			–	7.90
39.2	62.1	16.97	84.68			–	90.10
		16.97	82.64			–	88.12
		20.55	93.09			–	94.18
		35.93	138.33	7.8–8.1	9.3–9.5	–	25.34
		26.72	108.75			–	102.13
		1631.37	10424			–	13.36
122.		5.43	10.97			–	20.22
105		0.16	0.23			–	8.65
120		0.17	0.21			–	14.30
160		0.20	0.24			–	12.79

NAME	FORMULA	ρ_b, g/cc	Φ_{SNP}, pu	Φ_{CNL}, pu
SILICATES				
Quartz	SiO_2	2.64	–1	–2
β–Cristobalite	SiO_2	2.15	–2	–3
Opal (3.5% H_20)	$SiO_2\ (H_2O)_{.1209}$	2.13	2	2
Garnet	$Fe_3Al_2\ (SiO_4)_3$	4.31	3	7
Homblende	$Ca_2\ NaMg_2Fe_2AlSi_8O22(O,OH)_2$	3.20	4	8
Tourmaline	$NaMg_6Al_6B_3\ Si_6O_2(OH)_4$	3.02	16	22
Zircon	$ZrSiO_4$	4.50	–1	–3
CARBONATES				
Calcite	$CaCO_3$	2.71	0	–1
Dolomite	$CaCO_3MgCO_3$	2.88	2	1
Ankerite	$Ca(Mg,Fe)(CO_3)_2$	2.86	0	1
Siderite	$FeCO_3$	3.89	5	12
OXIDATES				
Hematite	Fe_2O_3	5.18	4	11
Magnetite	Fe_3O_4	5.08	3	9
Geothite	$FeO(OH)$	4.34	50+	60
Limonite	$FeO(OH)(H_2O)_{2.05}$	3.59	50+	60
Gibbsite	$Al(OH)_3$	2.49	50+	60
PHOSPHATES				
Hvdroxvapatite	$Ca_5(PO_4)3OH$	3.17	5	8
Chlorapatite	$Ca_5(PO_4)_3Cl$	3.18	–1	– 1
Fluorapatite	$Ca_5(PO_4)_3F$	3.21	–1	–2
Carbonapatite	$(Ca_5(PO_4)_3)_2CO_3H_20$	3.13	5	8
FELDSPARS–Alkali				
Orthoclase	$KAlSi_3O_8$	2.52	–2	–3
Anorthoclase	$KAlSi_3O_8$	2.59	–2	–2
Microcline	$KAlSi_3O_8$	2.53	–2	–3
FELDSPARS–Plagioclase				
Albite	$NaAlSi_3O_8$	2.59	–1	–2
Anorthite	$CaAl_2Si_2O_8$	2.74	–1	–2
MICAS				
Muscovite	$KAl_2(Si_3AlO_{10})(OH)_2$	2.82	12	20
Glauconite	$K_2(Mg, Fe)_2\ Al_6(Si_4O_{10})_3(OH)_2$	~2.54	~23	~38
Biotite	$K(Mg, Fe)_3(AlSi_3O_{10})(OH)_2$	~2.99	~11	~21
Phlogopite	$KMg_3(AlSi_3O_{10})(OH)_2$			

Table 6–1. *Logging Tool Response in Sedimentary Materials*

t_c µs/ft.	t_s µs/ft.	P_e barn/elect	U barn/cc	ε farads/m	t_p nsec/m	GR API units	Σ c.u.
56.0	88.0	1.81	4.79	4.65	7.2	–	4.26
		1.81	3.89			–	3.52
58		1.75	3.72			–	5.03
		11.09	47.80			–	44.91
43.8	81.5	5.99	19.17			–	18.12
		2.14	6.46			–	7449.82
		69.10	311				
49.0	88–4	5.08	13.77	7.5	9.1	–	7.08
44.0	72	3.14	9.00	6.8	8.7	–	4.70
		9.32	26.65			–	22.18
47		14.69	57.14	6.8–7.5	8.8–9.1	–	52.31
42.9	79.3	21.48	111.27			–	101.37
73		22.24	112.98			–	103.08
		19.02	82.55			–	85.37
56.9	102.6	13.00	46.67	9.9–10.9	10.5–11.0	–	71 12
		1.10				–	23.11
42		5.81	18.4			–	9.60
42		6.06	19.27			–	130.21
42		5.82	18.68			–	8.48
		5.58	17.47			–	9.09
69		2.86	7.21	4.4–6.0	7.0–8.2	–220	15.51
		2.86	7.41	4.4–6.0	7.0–8.2	–220	15–91
		2.86	7.24	4.4–6.0	7.0–8.2	–220	15.58
49	85	1.68	4.35	4.4–6.0	7.0–8.2	–	7.47
45		3.13	8.58	4.4–6.0	7.0–8.2	–	7.24
49	149	2.40	6.74	6.2–7.9	8.3–9.4	–270	16.85
		6.37	16.24				24.79
50.8	224	6.27	18.75	4.8–6.0	7.2–8.1	–275	29.83
50	207						33.3

7

Putting It All Together

Interpreting a log is a lot like trying to hug an elephant: first you have to get your arms around it. So much data about the well is collected that it is often difficult to assimilate, or get your arms around, everything you need to make the best interpretation. Unfortunately, many decisions are made on the basis of incomplete or neglected information.

QUESTIONS TO ASK BEFORE READING A LOG

Your first step is to compile as much information as possible about the well or prospect. What is the primary objective? Is there production from this zone in the area? What type of production: oil or gas? How much? What is the cumulative production from nearby wells? Are logs available from nearby wells? What is the lithology of the producing formation: sandstone, dolomite, or limestone? What are the porosity, resistivity, formation water resistivity, and water saturations from offset producing wells? What does the geological map look like? Are there any secondary objectives?

Answer these questions, preferably before you drill the well. Then keep the answers fresh in your mind when you examine the logs from your well. A good way to do this is to set up a table listing the various parameters from the offset wells.

In the case of a rank wildcat (a well drilled in an area that has never had a producing well), a different set of questions must be answered. These questions are based more on geologic and possibly seismic data. Estimates can be made for a best case/worst case scenario for porosity, resistivity, water saturation, and, ultimately, reserves. This practice usually provides the economic justification for drilling the well.

The second step is to get a feel for this well. Read the daily drilling reports. Note any unusual occurrences recorded, such as shows, kicks, lost circulation, or sticking. Next, study the mud log in detail. Again, look for indications of hydrocarbons in samples that represent good formation rocks, and note signs of porosity such as drilling breaks. Talk to the mud logger if possible. The logger can sometimes provide important details that are not included on the mud log.

After all this, you should have some kind of idea of what you'll see on the logs.

READING A LOG

Normally you will have a resistivity log and one or more porosity logs to look at. One of the logs, usually the resistivity log, will have been run on a correlation scale of 1 or 2 in./100 ft. In addition, the logs will have a detail scale of 5 in./100 ft. First compare the log that was run on the correlation scale with the same-scale log on the offset wells. Use the SP, GR, resistivity, and/or porosity curves to mark the formation tops as you identify them (this is called *making the correlation*). After identifying the main and the secondary objectives, if any, mark the correlation on all the logs. Ask yourself, "Are the formations in this well running higher or lower than expected? Are they higher or lower than the offset wells?" Unless a fault has been crossed, you want the well to run high because oil and gas are lighter than water and are found at the top of a zone.

Second, compare the porosity of the new well to the porosity of the offset well. Usually a minimum porosity is selected, called the *porosity cutoff.* (The porosity cutoff will vary with the different formations. It may be as low as 5% or as high as 15–20%.) Below the porosity cutoff the reservoir is too tight to produce; this part of the formation is not counted as *pay* (producible formation). How many feet of formation in the objective zone are above the porosity cutoff (this is called the *net pay*)? Is the porosity greater or less than the offset? Does the new well have more or less net pay than the offset? Is the reservoir quality as good as in the offset (is it shalier and poorer quality or cleaner and better quality)?

A good way to quantify reservoir quality is either to add the porosity, foot by foot, through the pay zone or to multiply the zone thickness (h) by the average crossplot porosity corrected for shale. The result is called *cumulative porosity-feet.* After the porosity cutoff is established, any formation with a porosity lower than the cutoff is not counted as net pay.

Third, ask yourself whether the resistivity is higher, lower, or equal to the offsets. Do any characteristic resistivity profiles suggest a gas/oil, gas/water, or oil/water contact? Is there anything on the mud log or in the known geology of the area that might suggest low-resistivity pay zones (resistivity curves affected by a low-resistivity mineral like pyrite)? Does the R_w calculated from the SP curve agree with the R_w used on the offset wells?

By examining the new well in this manner, you can decide whether the prospect should be better, worse, or about the same as the offsets. This kind of interpretive procedure takes a global approach. It assumes that nothing exists independently—that the formations are continuous and more or less alike in a given area and age. The approach is based on common sense.

What we've done so far is not a complete interpretation but a quick-look evaluation. It works very well when the new well has *offset production* (production from a nearby well). We need little expertise in log interpretation for this kind of evaluation. Naturally, more detailed methods of log interpretation are used even on development wells and must be used on wildcats where no offset production exists.

BVW_{min} QUICK-LOOK METHOD

Another quick-look interpretation procedure uses the concept of minimum bulk volume water (BVW_{min}). This theory states that the amount of water a formation can retain without producing any water is constant for a particular formation. Therefore, if we calculate BVW for a formation and it is less than or equal to BVW_{min}, the formation will produce free of water. (Water-free production is obviously desirable because it costs money to produce and dispose of the unwanted water.) The values of BVW_{min} are about

3.5% for a carbonate and from 5% for a clean sandstone to as high as 14% for a shaly sandstone.

The amount of water in the formation is equal to water saturation times porosity (BVW = $S_w \times \Phi$). Since S_w and Φ are two of the terms in the Archie equation (see Chapter 3), let's transpose and substitute terms so we can see the significance of BVW_{min}:

$$S_w^2 = R_w/(\Phi^2 \times R_t)$$

Let's see what happens if we rearrange this equation in terms of BVW:

$$S_w^2 \times \Phi^2 = (S_w \times \Phi)^2 = BVW^2 = R_w/R_t$$

$$BVW_{min}^2 = R_w/R_{tmin}$$

$$R_{tmin} = R_w/(BVW_{min})^2$$

where:

R_{tmin} = minimum formation resistivity needed for water-free production

If we know the formation water resistivity and BVW_{min}, we can calculate the approximate R_{tmin} needed to ensure water-free production. For a carbonate we need about 800 x R_w; for a sandstone we need about 200 (slightly shaly) to 400 (clean) times R_w. This technique, a member of the "quick-and-dirty" school of log interpretation, seems to work best in hard-rock country where matrix density (ρ_{md}) is 2.68 for sandstones.

Once we have identified the zones that have enough resistivity to produce water free, we only need to discover whether the porosity is high enough. For an unfractured carbonate, we need at least 8% crossplot porosity (10% for a sandstone). These are shale-corrected crossplot porosities.

The best way to apply this technique is to draw cutoff values for resistivity and porosity on the logs. First, we determine the R_t cutoff (equal to R_{tmin}) and draw a line down the dual induction log or dual laterolog at that resistivity. Any zone with the deep resistivity curve reading to the right of the cutoff (higher resistivity) is a possible candidate for production. Next, we draw a porosity cutoff on the porosity log. Porosity cutoffs are dependent on the type of reservoir, the geologic area, and the geologist working the area.

In low-porosity areas (hard-rock country), 8% for a carbonate and 10% for a sandstone are common cutoffs. In high-porosity areas, higher values may be used. Any zones with a crossplot porosity to the left of the porosity cutoff (higher porosity) and a resistivity to the right of the resistivity cutoff should be productive.

Normally we use either the crossplot porosity or the density-porosity. (Make any necessary adjustments for changes in lithology if you use only the density porosity.) We also note any gas effects on the neutron-density porosity log. (Caution: Be sure any separation on the neutron density log is not due to a lithology change. If in doubt, read Chapter 6.)

This technique is a good scouting device that will indicate where to take a closer look. Once you have used this technique a couple of times, it will be easier and faster than reading about it. Use this method and all "quick-and-dirty" techniques with caution. They are useful as a first look. Always conduct a detailed analysis before making important, costly decisions.

SAMPLE READING

Let's say your Uncle Howard, the oilman, has just drilled and logged the Sargeant 1–5 well in the lower Cunningham formation. Uncle Howard is giving you a once-in-a-lifetime chance to invest in (he says) a sure thing. To prove that he has always liked you best, he will let you see the logs before you decide whether to invest your life savings in this sure thing. Unc will give you all the information he has, but you have to make up your own mind.

On the basis of what you've learned so far about interpretation, start collecting the information you need to make your decision. Figure 7–1 is a production map for nearby wells. Locate the Sargeant 1–5. You can see a 10-bcf (billion cu ft. of gas) well, the Nora 1–32, about a mile to the north. (Each square on the map represents a square mile.) The Corporal 1–33, one-half mile to the northeast, had initial production (IP) of 2 MMcfd (million cu ft. of gas/day) and reserves of 2 bcf. The nearby Private 1–5 has a cumulative production of 35 MMcf. (Unfortunately, you don't have a reserve estimate or IP.) Other apparently productive wells lie to the west and northwest. The nearest dry hole is the Sam 1–1, nearly 2 miles to the west–southwest. So far, things look good.

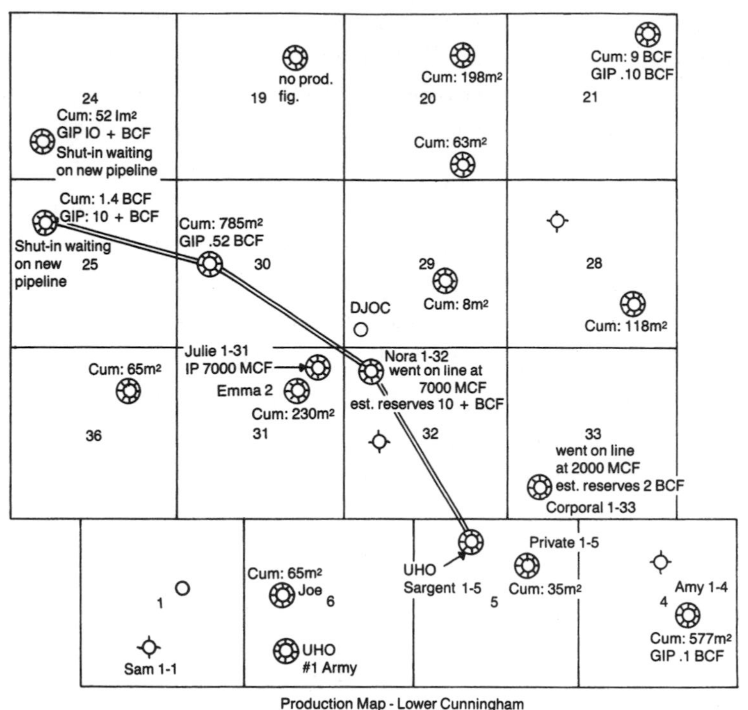

Fig. 7–1. *Production map for the lower Cunningham, showing offset wells and their production.*

Now look at Figure 7–2, the *net isopach* or geologic map. It shows the net zone thickness using an 8% porosity cutoff. That is, the geologist counted only that part of the reservoir that was ≥ 8% porosity. The geologist then drew the isopach (equal thickness) map.

The lower Cunningham is one of the Springer series of sandstones. Each contour line represents a change in thickness of 5 ft. Zone thickness increases from 0 at the edge to a maximum of 20 ft. Note that the Nora 1–32 has 18 ft. of pay out of a gross interval of 39 ft. (18/39). The Julie 1–31 that also came in at 7 MMcfd has 7 ft. of net pay. The Emma 2–31 has 0 ft. net pay with 10 ft. of gross pay; the production map (Fig. 7–1) shows a cumulative production of 230 MMcf during an unknown period. Check the net and gross feet of pay on the surrounding wells. Note that your prospect, the Sargeant 1–5, is estimated to have 6 ft. of net pay.

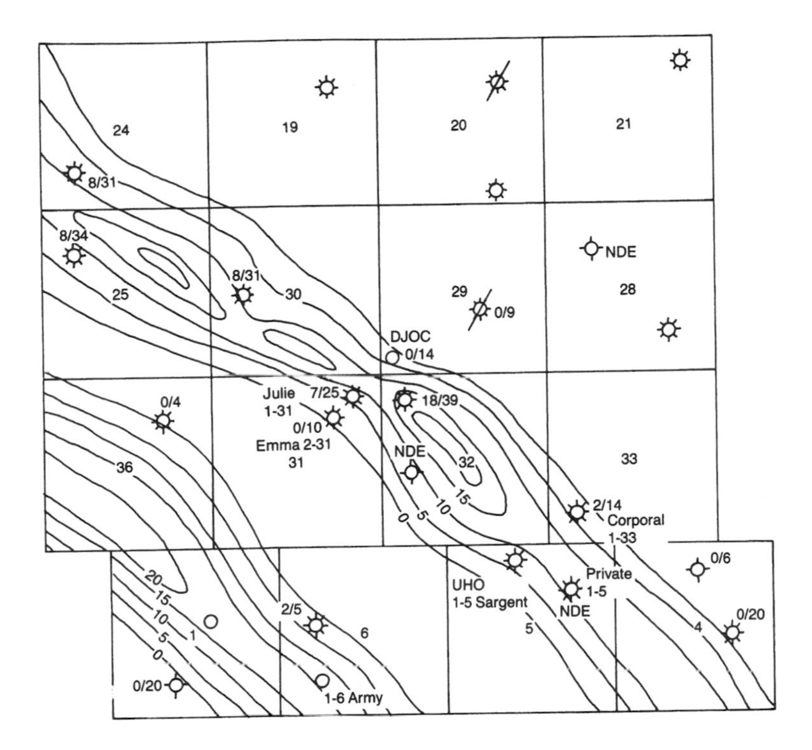

Fig. 7–2. *Net isopach, or geologic, map of the lower Cunningham.*

You're starting to amass a lot of information, so this would be a good time to set up a table to keep track of it.

Well Name	h	Reserves	IP	Cumulative Production
Nora	18	10	7	–
Corporal	2	2	2	–
Private	11	–	–	35
Julie	7	–	7	–
Emma	0	–	–	230
Sargeant	6	?	?	–

As you sift through the information that Uncle Howard furnished, you complete the table, shown here as Table 7–1.

Well	h	Φ	h x Φ	R_w	R_t	S_w	BVW	Remarks
Lower Cunningham								
Nora	18	20	360	0.12	75	20	0.040	10 bcf, 7MM IP
Corporal	2	15	30	0.12	68	28	0.042	2 bcf, 2MM IP
Emma	0	7	0	0.14	52	68	0.048	230MM cum
Julie	7	–	–	–	65	–	–	7MM IP
Private	11	–	–	0.12	–	50	–	35 MM cum
Sargeant								
Morrow								
Army	26	12	312	0.22	60	50	0.06	250M/15-BWPD
Private	22	10	220	0.2	42	72	0.07	Not tested

Table 7–1. *Data on Lower Cunningham and Morrow*

Next, look at the mud log (Plate 1, inserted at the back of the book). Was there a show when the driller cut through the Morrow or the lower Cunningham? The formations are marked on the log, and we can see on the Morrow a slight show (135 units on the hot-wire gas detector; 160 units C_1 on the gas chromatograph). Background gas did not increase after drilling through the zone. On the lower Cunningham from 12,732 to 12,755 ft., the mud log shows 445 units on the hot-wire detector and 300 units on the GC. The formation is described as sandstone, white to light gray, fine grained, friable, subangular, fair sorting, slightly calcareous, no fluorescence, faint dull yellow ring cut. Below 12,755 ft., background gas has increased from 60 to 120 units. So from the mud log we can conclude that the lower Cunningham is promising. The results are not conclusive, however, for either zone.

Even before you look at the wireline logs on the Sargeant, you can set up some criteria on which to base your evaluation. Since the lower Cunningham is a sandstone reservoir, you'd like to have at least 10% porosity. However, the Emma has made 230 MMcfg with only 7% porosity, and the geologist mapped the formations on an 8% porosity cutoff, so use 8% as your minimum porosity. For a slightly shaly sandstone, R_t should be at least 200 times R_w to produce water-free. According to Table 7–1, R_w is about 0.12, so you need 200 x 0.12 = 24 ohms in the lower Cunningham. (We use the lowest R_w from the offsets because we don't want to eliminate any potential production in our scouting method. We'll refine the zones later.)

How about the Morrow? What should you use for a porosity and resistivity cutoff there? Notice that the R_w in the Morrow seems to be much fresher (higher resistivity) than in the lower Cunningham. (Check the dual induction log in Plate 2 at the back of the book and the compensated neutron density log in Plate 3 where the cutoffs are already drawn.)

Since the lower Cunningham is the primary objective, look at it first. Draw the resistivity cutoff at 24 ohms on the dual induction log from the top of the lower Cunningham to the bottom of the log. To make the zones stand out, many engineers and geologists color the separation between the deep induction curve and the resistivity cutoff. Here you can see four intervals where the resistivity is higher than the cutoff, labeled from 1–4.

Transfer the intervals identified on the resistivity log to the porosity log. Draw a small rectangle at each zone, 1–4. Next, draw the cutoff at 8% porosity through the primary objective. Now you must approximate a crossplot porosity calculation. Use the average of the neutron and density porosity curves, and mark this porosity at each of the intervals, 1–4. (The average is generally very close to the crossplot porosity that is obtained from the neutron density chart as long as the zone is not too shaly. The average is fine for a first approximation.) Darken any of the rectangles whose porosity is 8% or greater. Only zone 2 (and not all of it) has a crossplot porosity high enough to be darkened. Zone 1 is about 4.5%, zone 3 is 6.5%, and zone 4 is 4%. Zone 2 has 8.5% porosity and should be productive according to the quick-and-dirty technique.

Now look at the Morrow zone. We'll use the same approach. Remember R_w is at least 0.2, so draw the resistivity cutoff at 40 ohms (200 x 0.2). The lower Morrow is divided into intervals 5, 6, and 7 for easier study. The interval above the shale break is zone 8. (These zones are marked on the log in Plate 3.) Transfer the zones from the resistivity log to the porosity log and then draw the crossplot porosity on the log (you can draw Φ_{xp} by eye accurately enough for this step, or add Φ_N to Φ_D and divide by 2). Only part of interval 8 has a crossplot porosity > 8%. It should also be productive.

What do you think? Will you invest the kids' college fund in this one?

Don't be hasty. The quick-and-dirty technique is just a scouting device; it points out the zones of interest so you can look at them in more detail. Go back to the table that listed some of the parameters from the offset wells and see how the Sargeant stacks up (Table 7–2). Add whatever information you can.

Well	h	Φ	h x Φ	R_w	R_t	S_w	BVW	Remarks
Lower Cunningham								
Nora	18	20	360	0.12	75	20	0.040	10 bcf, 7MM IP
Corporal	2	15	30	0.12	68	28	0.042	2 bcf, 2MM IP
Emma	0	7	0	0.14	52	68	0.048	230MM cum
Julie	7	-	-	-	65	-	-	7MM IP
Private	11	-	-	0.12	-	50	-	35MM cum
Sargeant	7	8.5	60	-	60	-	-	Probably gas, tight

Table 7–2. *Additional data on Lower Cunningham and Morrow*

You see that the porosity is low but the porosity-feet are twice as high as the Corporal, which has reserves of 2 bcf. Also, the Emma has production with only 7% porosity. The Corporal had an IP of 2 MMcfd with 15% porosity. Since you have only 8.5%, the permeability (the ease with which the well will flow) is likely to be much less.

Before you make a final decision about whether to put your money into this well, estimate the gas reserves. You can make a rough estimate by using BVW_{min} once more. The equation for calculating gas reserves (Gp) is

$$
\begin{aligned}
G_p &= \text{Porosity} \times \text{Gas Saturation} \times \text{Thickness} \times \text{Area} \times \\
&\quad (\text{Initial Gas Expansion Factor} - \text{Final Gas Expansion Factor}) \\
&\quad \times \text{Conversion Constant} \\
&= F \times (1 - S_w) \times h \times A \times [(1/B_{gi}) - (1/B_{gf})] \times 43{,}560 \text{ ft.}^3
\end{aligned}
$$

where:

$(1 - S_w)$ = gas saturation, S_g

A = drainage area of reservoir, acres

h = feet of pay

$1/B_{gi}$ = initial gas expansion factor = 275

$1/B_{gf}$ = final gas expansion factor = 50

Conversion constant = 43,560 ft._2/acre

Let

$$1/B_g = (1/B_{gi}) - (1/B_{gf})$$

Rearranging and substituting BVW for $S_w \times \Phi$, you get

$$G_p = (\Phi - BVW) \times h \times A \times 1/B_g \times 43{,}560$$

Zone 2:

$$G_p = (0.085 - 0.05) \times 7 \times 640 \times 225 \times 43{,}560 = 1.54 \text{ bcf}$$

Zone 3:

$$G_p = (0.065 - 0.05) \times 5 \times 640 \times 225 \times 43{,}560 = 0.5 \text{ bcf}$$

The estimated reserves are 2.0 bcf for the lower Cunningham, your primary objective. The probable gas price is $1.25/Mcf, so you could anticipate a total sales price of $1.25 x 2,000,000 Mcf (note that 1 Mcf = 1000 ft.[3]) = $2,500,000. The net revenue interest is 78% (that is, of $1.00 in sales, 22¢ goes to royalty owners and override interests) less 7.085% severance tax. So the net income will be $1,770,000. The cost of drilling and completing the well will be $1,200,000.

If you assume that the Morrow is productive, you could have a G_p of 2.9 bcf. Net income from the Morrow could be as high as $2,500,000. However, there is no Morrow production in the area, so you face a good chance of no income from the Morrow. This is the way things look:

	Best Case	**Most Likely**	**Worst Case**
Well cost	($1,200,000)	($1,200,000)	($1,200,000)
Net income LC	1,782,000	1,782,000	1,215,000
Net income Morrow	2,600,000	1,300,000	0
Add'l $ to complete Morrow	(150,000)	(150,000)	(150,000)
Profit	$3,032,000	$1,732,000	($135,000)
Return on investment	2.24	1.28	loss

Your uncle wants to cut you in for 1% of this deal (about $13,500). What do you say?

The most sensible route would be to thank your uncle and decline. With the amount of risk involved, you should get a better payout than a best case of 2.24. (Remember that a well takes several years to produce its gas.) The most likely case is that you might break even; you could certainly lose some or all of your money. But if you're still interested in the well, you need to fine-tune your calculations so you're using the best numbers possible for your decision.

In Chapter 8, when you learn how to complete a detailed interpretation, we'll look again at Uncle Howard's well, and you'll see just how close you came with the quick-and-dirty technique. Naturally, if you are investing your own or someone else's money in a well, you would want a detailed analysis. Since this requires a certain amount of expertise, we recommend you seek someone who specializes in log interpretation—either someone in your own company, someone in one of the logging companies (usually their opinions are free), or a consultant.

8

Detailed Interpretation

Most log interpretation, at least in the basic sense of trying to determine whether a well will produce hydrocarbons, is involved with solving the Archie equation:

$$S_w^n = (F_R \times R_w) / R_t$$

where:

S_w = water saturation
F_R = formation resistivity factor
R_w = formation water resistivity
R_t = true formation resistivity
n = saturation exponent

The formation resistivity factor is related to the formation porosity according to

$$F_R = K_R / \Phi^m$$

where:

K_R = constant
m = cementation exponent; may vary from about 1.6 to 2.2

Common equations in use relating FR to Φ are the following:

$F_R = 1/\Phi^2$ (hard-rock country)
$F_R = 0.62/\Phi^{2.15}$ (Humble equation for high porosities)
$F_R = 0.81/\Phi^2$ (high porosities)

To solve the Archie equation, you need to know R_w, crossplot porosity corrected for shale (Φ_{xp}), and R_t. You can obtain R_w from water samples in offset wells, from calculations using porosity and the resistivity of an obviously 100% wet formation in the well, or from the SP curve. Porosity can be found with one or more of the various porosity-measuring devices at your disposal.

Normally you need more than one porosity tool to make a good estimate of porosity. You can usually assume a relationship between F_R and porosity based on the expected range of porosities. R_t is determined from one of the resistivity-measuring devices. Often the deep induction or deep laterolog curve is used as R_t without correction. If invasion is deep or the zones are thin (< 10 ft.), the resistivity deep curves must be corrected.

Once you have this information on a foot-by-foot or zone-by-zone basis (if you want average values), calculate S_w. With S_w as one piece of your total information on the well, decide whether to run pipe, test further, or plug and abandon the well.

You can use the porosity, water saturation, and formation thickness values; an estimate of the area that the well is expected to drain; and the expected type of production to make a hydrocarbon-in-place calculation from the following equations:

For oil: $N = (1 - S_w) \times \Phi \times A \times h \times B_o \times 7758$ bbl
For gas: $G = (1 - S_w) \times \Phi \times A \times h \times 1/B_g \times 43{,}560$ ft.^{3}

where:

N = total oil in place
G = total gas in place
A = drainage area, acres
h = net pay thickness, ft.
B_o = oil shrinkage factor, a measurement of how oil shrinks as it reaches surface and temperature and pressure
$1/B_g$ = gas formation volume factor, a measurement of how gas expands when it reaches surface temperature and pressure
7758 = constant to convert acre-ft. to bbl
43,560 = constant to convert acre-ft. to ft.^{3}

To convert oil or gas in place to a reserve number (the amount that can actually be recovered), multiply by the recovery factor. For oil, the recovery factor ranges from 0.05–0.90; 0.4 is normal. Gas recovery depends on the abandonment pressure and the type of reservoir drive; it is usually 0.7–0.9 for a water drive. For an expansion gas drive, you can use the difference between $1/B_{gi}$ (initial) and $1/B_{ga}$ (at abandonment).

Let's work through a couple of log interpretations and see how to come up with the various parameters you need to solve for water saturation. Start with Uncle Howard's well from Chapter 7 and see how accurate your preliminary interpretation was.

SARGEANT 1–5 EXAMPLE

You'll be working with the dual induction log (Plate 4) and the compensated neutron-density porosity log (Plate 5) located at the end of the book. Both logs have GR curves, the dual induction has an SP curve, and the porosity log has a caliper curve, all in track 1.

Usually the first thing to do is determine or check the R_w valuc. We used 0.12 ohms for the lower Cunningham and 0.20 ohms for the Morrow on the basis of information from the offsets and local knowledge. However, always verify R_w if possible by one or both of the following methods.

First, look for an obviously wet sand. If you can identify one by its low resistivity, you can calculate R_w from the equation

$$R_o = F_R \times R_w$$

$$R_w = R_o / F_R = R_o / (1/\Phi^2)$$

$$= R_o \times \Phi^2 \text{ (for low } \Phi \text{ areas)}$$

Looking over the lower Cunningham, you note that nothing is obviously wet. [The ability to recognize wet (S_w = 100%) zones comes with experience. Wet zones usually have low resistivities and low GR readings.] The low resistivity readings on this well are associated with shales; you can tell this from the high GR readings and from the porosity logs. The neutron curve reads high porosity, and the density reads low porosity in the shales.

On the Morrow section, one zone at about 12,640 ft. has a clean GR log and 80 ohms resistivity. The crossplot porosity is 5.5%. To check R_w,

$$R_w = (\Phi^2 \times R_o)$$
$$= (0.055)^2 \times 80 = 0.24$$

This is close to your assumed value for R_w, but let's make another check.

The next choice for determining or checking R_w is the SP curve. The deflection of the SP is proportional to R_w and R_{mf} (the mud filtrate resistivity). Unfortunately, there is little or no movement on the SP curve in the lower Cunningham—often the case in low-porosity, low-permeability formations. (Also, the SP will show no deflection opposite a permeable zone if R_{mf} and R_w are the same. If $R_{mf} < R_w$, the SP will deflect to the right toward the depth track.) The upper part of the Morrow shows a small deflection of about 0.7 divisions opposite the best porosity. Let's check R_w here:

$$SP_{mv} = -K_c \log (R_{mfe}/R_{we})$$

where:

K_c = constant that varies with temperature
R_{mfe} = equivalent mud filtrate resistivity
R_{we} = equivalent formation water resistivity

The SP equation is correctly written in terms of the ratio of the chemical activities of two sodium chloride solutions rather than their resistivities. Because the resistivity of the formation water and the mud filtrate is due to other factors in addition to sodium chloride, each resistivity must be converted to a sodium chloride equivalent resistivity. Do this by using Figure 8–1.

To calculate R_w, you need the SP reading (0.7 div × (–20) mv/div = –14 mv); R_{mf} from the log heading (1.08 @ 69 °F); and BHT (bottom-hole temperature) from the log heading (213 °F). Next, convert R_{mf} at measured temperature to R_{mf} at bottom-hole temperature with the general equation

$$R_2 = R_1 \times [(T_1 + 7)/(T_2 + 7)]$$

$$R_{mf} = 1.08\ [(69 + 7)/(213 + 7)] = 1.08\ (76/220)$$

$$= 1.08\ (76/220) = 0.37 \text{ ohms at } 213\ ^\circ\text{F}$$

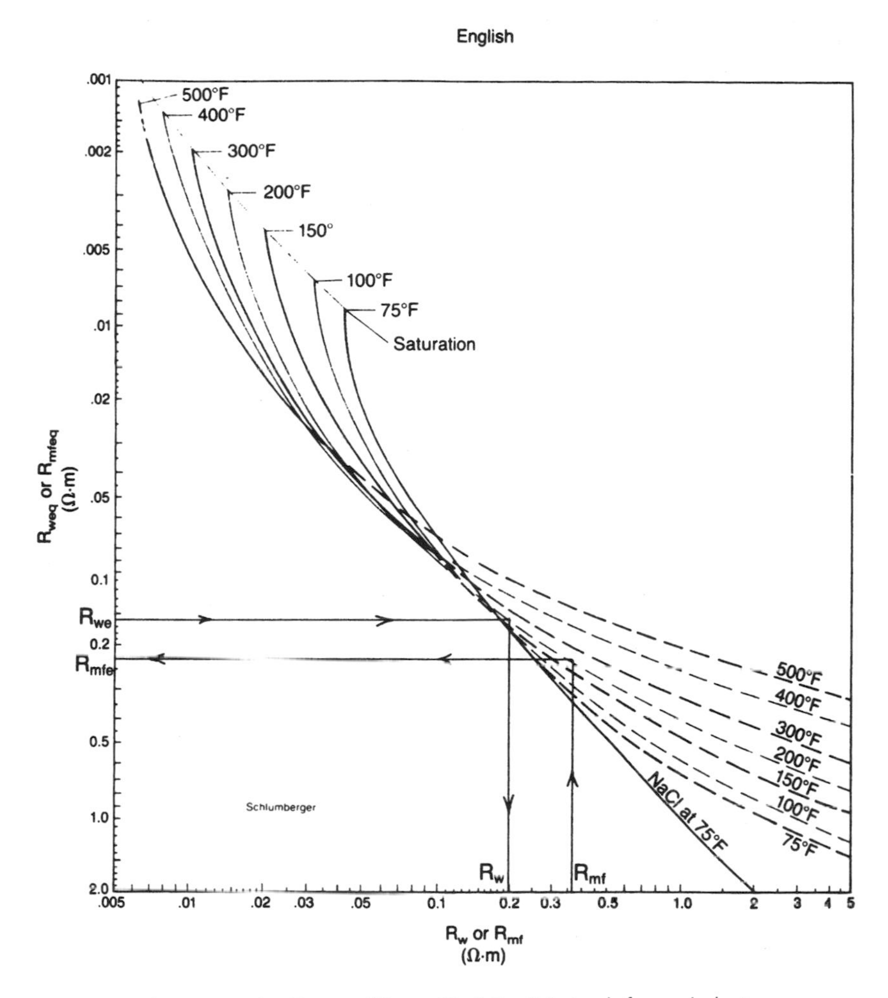

Fig. 8–1. *Chart converting R_{we} and R_{mfe}. The letter "e" stands for equivalent (courtesy Schlumberger).*

Convert R_{mf} at BHT to R_{mfe} using Figure 8–1. Enter the bottom scale with R_{mf}= 0.37. Draw a vertical line until it crosses the 200 °F line. Eyeball where 213 °F is between the 200 °F and 300 °F lines. From this point, draw a horizontal line to the left edge and read R_{mfe} = 0.22.

On Figure 8–2 find the SP reading at the bottom and draw a vertical line until it intersects the 200 °F temperature line. From this intersection draw a horizontal line to the right to determine the R_{mfe}/R_{we} ratio. For SP = –14 mv and 200 °F, R_{mfe}/R_{we} = 1.4. Continue with the nomograph (a nomograph is a graphical method of solving an equation) and solve for R_{we} by drawing a line from the R_{mfe}/R_{we} point through the R_{mfe} point on line 2 until it crosses line 3. This intersection is R_{we}. We read R_{we} = 0.16.

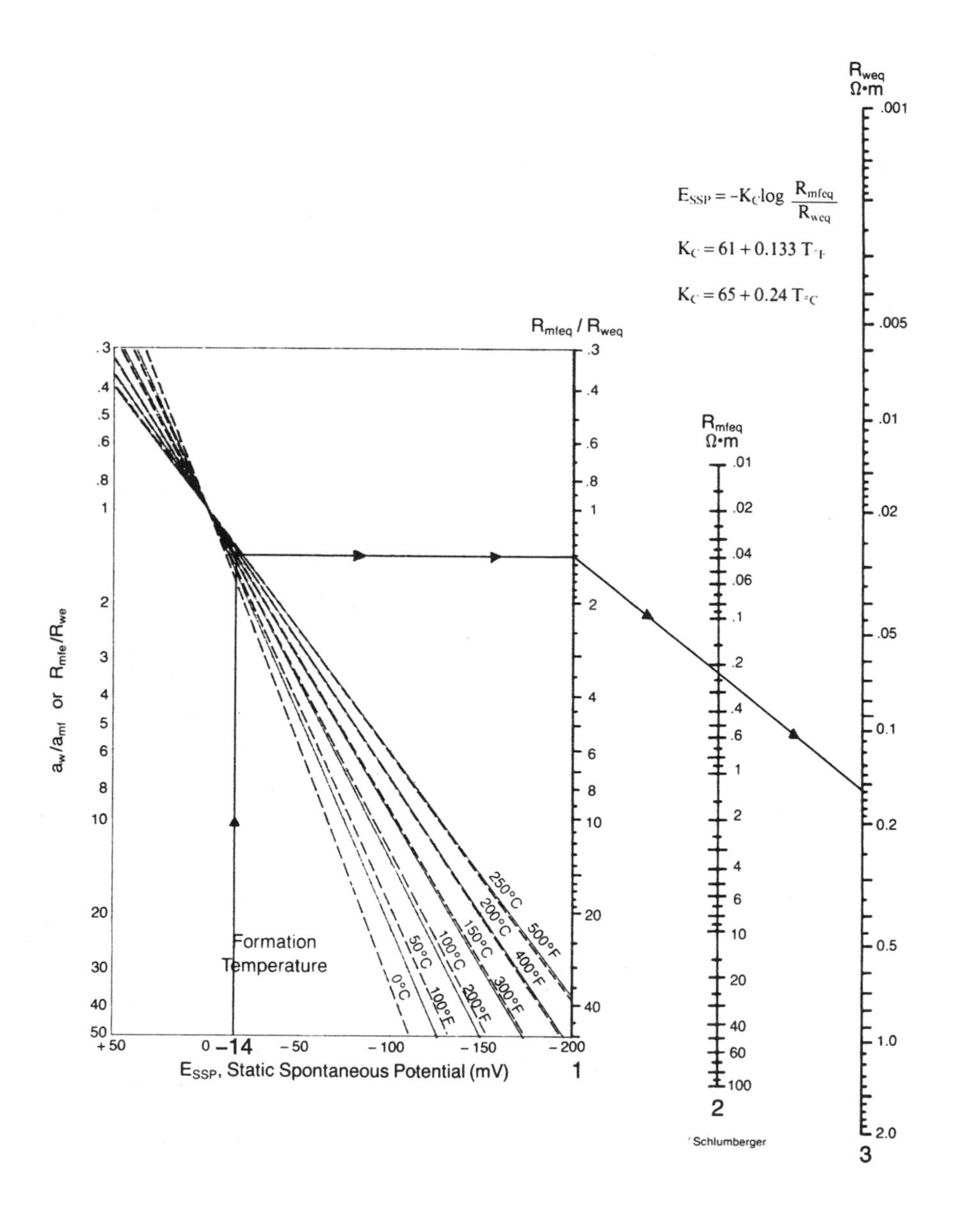

Fig. 8–2. *Nomograph for determining R_{we} from the SP log (courtesy Schlumberger).*

Now you must convert R_{we} to R_w. Start at the left edge of Figure 8–1 with R_{we} = 0.16. After intersecting the 213 °F line and dropping down to the bottom line, you should come out with R_w = 0.2. This value checks nicely with the information from your offsets, so use it for your Morrow calculations.

You weren't able to calculate R_w for the lower Cunningham, so use 0.12 because this is the best information available. You were able to verify that R_w = 0.2 in the Morrow from the SP.

The next step is to set up a log analysis table, as shown in Table 8–1. The Sargeant has already been analyzed. Note that the table is divided into two sections: one for the lower Cunningham and one for the Morrow. Also note that the first point is a shale zone. You need the shale information to determine V_{sh} and to correct the crossplot porosity to effective porosity.

When you're working with low-porosity formations, porosity is usually the most critical input. A zone can't produce anything—even water—if it doesn't have sufficient porosity. Therefore, look at the porosity logs first and record GR, Φ_D, and Φ_N for each zone. When using average values for a zone, as is done here, you can see the picks are subjective. Someone else might use slightly different numbers. Try to pick an optimistic rather than a pessimistic value. That way, if the logs condemn a zone as wet or tight, you can be sure it is.

The next step is to calculate V_{sh}. Note the shale baseline drawn on the porosity log. The lower Cunningham baseline is at 6.0 divisions on track 1. The clean sand line is drawn through the lowest GR reading. GR_{cl} (clean gamma ray) is usually about 1 division, as in this case. Now calculate V_{sh} in the lower Cunningham for zones 1 and 2:

Zone 1: $V_{sh} = (GR_{z1} - GR_{cl})/(GR_{sh} - GR_{cl})$

$= (1.8 - 1.0)/(6.0 - 1.0) = 0.8/5.0 = 0.16$

Zone 2: $V_{sh} = (1.0 - 1.0)/(6.0 - 1.0) = 0/5 = 0$

where:

GR_{z1} = gamma ray reading, zone 1

GR_{cl} = gamma ray reading, clean

GR_{sh} = gamma ray reading, shale

Table 8–1. *Data logged*

COMPANY UHOC											WELL Sargeant 1-5		
FIELD Evans Ranch					COUNTY						STATE		
DEPTH	GR	V_{sh}	R_S	R_M	R_D	R_{tmin}	Φ_D	Φ_N	Φ_{xp}	Φ_e	% SW	% BVW	REMARKS
12734-12748	6.0	1.0	3	2	2	2	16.2	30	23	–	–	–	Lwr. Cunn., Shale Pt.
12774-12784	1.8	0.16	1200	70	52	380	7.5	3.0	5.2	2.0	88	.018	Zn1 Tight
12754-12761	1.0	0	700	110	85	220	11.0	5.0	9.0	9.0	26	.023	Zn2 Gas
12739-12744	1.1	0.02	1500	350	85	480	12.0	2.0	7.0	6.5	24	.016	Zn3 Gas
12694-12698	3.5	0.50	400	40	40	130	6.0	2.0	4.0	0	–	–	Zn4 Tight
Lower Cunningham—R_w = 0.12 from offset wells													
12650-12660	5.0	1.0	4.5	3	3	3	9.0	25	17	–	–	–	Morrow Shale Pt.
12640	1.0	0	380	90	80	80	9.0	2.0	5.5	5.5	91	.05	Tight
12626-12634	1.8	0.2	900	180	130	120	9.5	5.5	7.5	4.0	100	.04	Zn5 Wet
12616-12622	2.3	0.32	200	90	160	160	8.5	3.0	6.0	.5	–	–	Zn6 Tight
12604-12611	1.7	0.17	300	450	120	120	9.0	3.0	6.0	3.0	100	.03	Zn7 Tight
12592-12598	1.0	0	350	120	200	200	10.0	3.5	7.0	7.0	45	.03	Zn8L Tight Gas
12583-12592	1.0	0	200	90	90	90	14.0	3.0	9.0	9.0	52	.047	Zn8U Gas

Morrow—R_w = 0.20 from SP

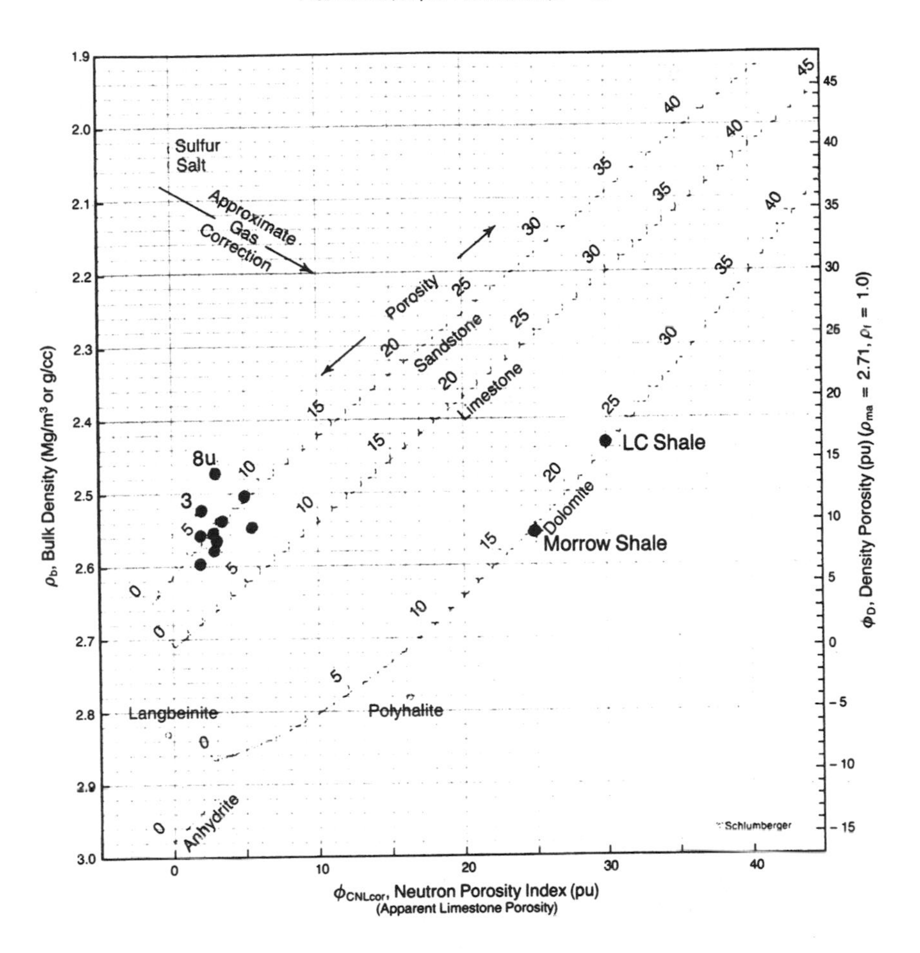

Fig. 8–3. *Crossplot porosity chart for a neutron density log (courtesy Schlumberger).*

When you come to the Morrow zone, check the shale and clean sand picks. Note that GR_{sh} is now 5.0, but GR_{cl} is still 1.0 in the Morrow.

Next, find the crossplot porosity and effective porosity for each zone. On Figure 8–3, find the density porosity on the right-hand scale and the neutron porosity on the bottom scale. Notice that the shale point (Φ_D = 16.2, Φ_N = 30) plots on the dolomite matrix line. This is where shale plots; it does not mean the zone is dolomite. The crossplot porosity for the lower Cunningham shale is 23%.

The crossplot porosity for zone 1 is 5.2 (Φ_D = 7.5, Φ_N = 3.0). Since some shale is in the formation as evidenced by V_{sh} = 0.16, we must correct the crossplot porosity to get effective porosity (Φ_e):

Zone 1:

$$\Phi_e = \Phi_{xp} - (V_{sh} \times V_{shxp})$$
$$= 5.2 - (0.16 \times 23) = 5.2 - 3.68$$
$$= 1.52 = 2 \text{ (after rounding)}$$

Zone 2:

$$\Phi_{xp} = 9.0 \ (\Phi_D = 12.0, \Phi_N = 5.0)$$
$$\Phi_e = \Phi_{xp} \text{ since } V_{sh} = 0$$

After calculating all V_{sh} and Φ_e, we determine R_t for the various zones. If the porosity is < 5%, we can skip the zone as being too tight to produce (unless it is naturally fractured).

Now enter the resistivity values for the shallow, medium, and deep curves in the table. If the formation is < 30 ft. thick, the resistivity values will have to be corrected for the effect of the surrounding shales. The corrected resistivity values are then entered in the proper tornado chart where we determine R_t.

The R_t values are already entered on Table 8–1. If you must make this kind of detailed correction, you will need a chart book from the appropriate logging company with its correction charts. Instructions for use are included in the chart books.

You can make one bed thickness correction that does not require correction charts by using the R_{tmin} method. If you wish to use this method, the bed thickness should be 10–25 ft. and the deep resistivity should be at least 10 times the shale resistivity. To find R_{tmin},

$$R_{tmin} = R_s \times (R_w / R_{mf})$$

where:

R_s = shallow resistivity reading

This method will give a zone's minimum theoretical resistivity based on the shallow resistivity reading and the invaded zone. You can apply this method to the lower Cunningham. If R_w = 0.12 and R_{mf} = 0.37, then

Zone 1: $R_{tmin} = 1200 \times (0.12/0.37) = 380$

Zone 2: $R_{tmin} = 700 \times (0.12/0.37) = 220$

For the Morrow, the zone is about 50 ft. thick, so bed thickness corrections are not necessary. The R_{SFL}, R_{IM}, and R_{ID} values taken from the resistivity log are entered in the tornado chart (Fig. 8–4), and R_t is calculated from the ratio R_t / R_d. Now calculate zone 5 (12,626–12,634 ft.). To use the tornado chart in Figure 8–4, you need the ratios R_{SFL} / R_{ID} and R_{IM} / R_{ID}. From Table 8–1 you have R_{SFL} = 900, R_{IM} = 180, and R_{ID} = 130. Then

$$R_{SFL}/R_{ID} = 900/130 = 6.92$$

$$R_{IM}/R_{ID} = 180/130 = 1.38$$

From the tornado chart,

$$R_t/R_{ID} = 0.90$$

$$R_t = (R_t/R_{ID} \times R_{ID})$$

$$= 0.90 \times 130 = 120$$

Now that you have R_t, R_w, and Φ_e, you can calculate S_w using the Archie equation. Finally, multiply S_w by Φ_e to find bulk volume water (BVW) and enter these two values for each zone in Table 8–1. Start with the lower Cunningham.

Zone 1: $\Phi_e = 2.0$

Zone 1 is too tight to producc.

Zone 2: $R_t = 220$

$$\Phi_e = 9.0$$

$$S_w = \sqrt{R_w/(\Phi^2 \times R_t)}$$

$$= \sqrt{.012/((0.09^2) \times 220)} = \sqrt{.012/1.782}$$

$$S_w = \sqrt{.673} = 0.26 = 26\%$$

$$BVW = S_w \times \Phi_e = 0.26 \times 0.09 = 0.023$$

Zone 2 will produce gas and no water.

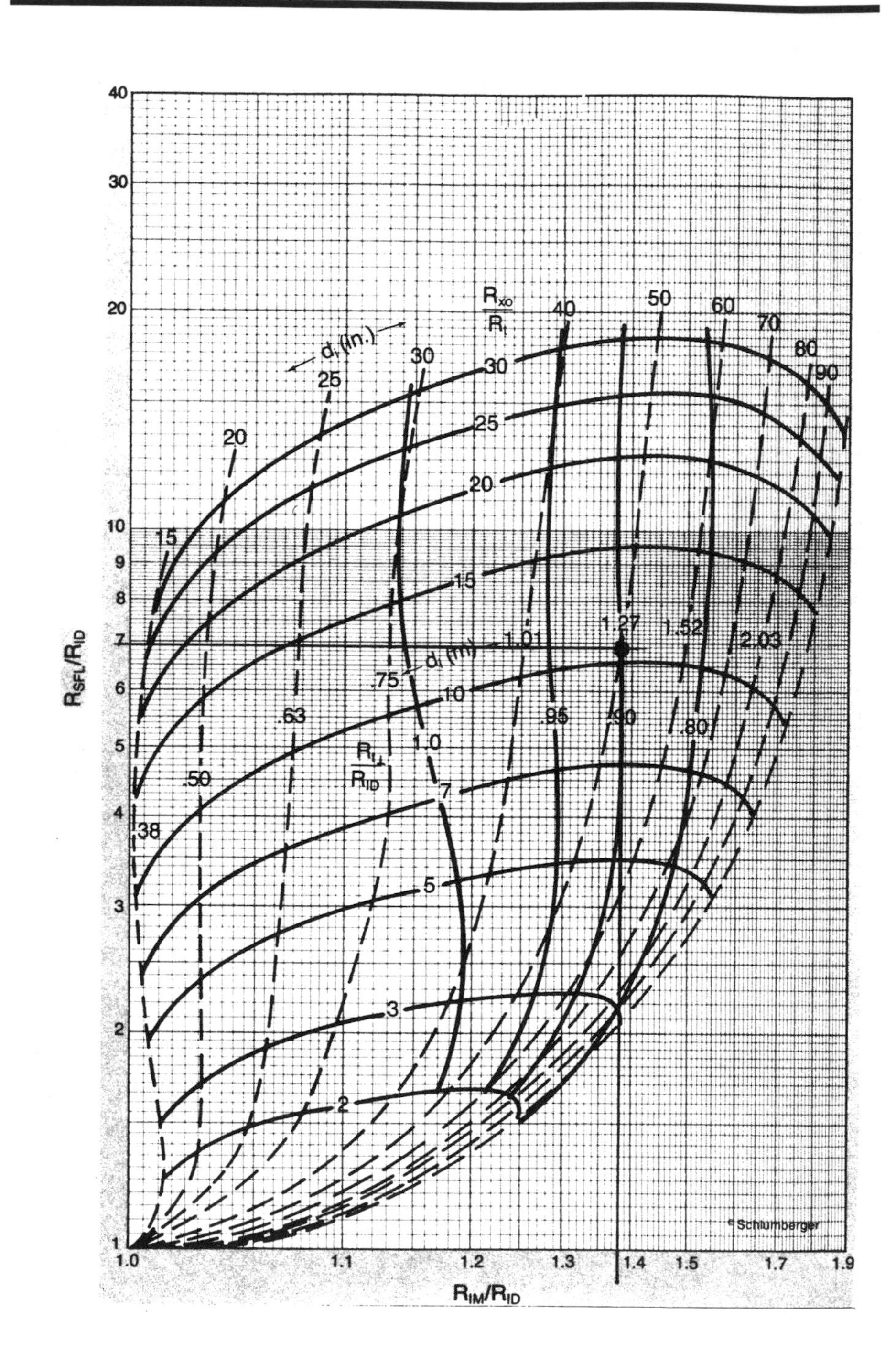

Fig. 8–4. *Tornado chart for a dual induction–spherically focused log (courtesy Schlumberger).*

Zone 3:

$$R_t = 480$$

$$\Phi_e = 6.5$$

$$S_w = \sqrt{.012/((0.665)^2 \times 480}$$

$$= \sqrt{.012/2.028}$$

$$= 0.24 = 24\%$$

$$BVW = 0.24 \times 0.065 = 0.016$$

If zone 3 produces anything, it will be gas with no water, but the porosity is very low.

Zone 4: $\Phi_e = 0$

No production.

In the Morrow, only zone 8 appears to be productive. Here S_w and BVW are both good. The zone should produce water-free gas.

The only task left is to complete a reserve estimate to see whether enough hydrocarbons are present to complete a commercial well. In the following equation, the subscripts *i* and *a* stand for initial and abandonment pressure, respectively. Also assume $(1/B_{gi}) - (1/B_{ga}) = 225$. The equation for gas reserves (G_p) is

$$G_p = 43{,}560 \times (1 - S_w) \times h \times \Phi \times A \times [(1/B_{gi}) - (1/B_{ga})]$$

Zone 2:

$$G_p = 43{,}560 \times (1 - 0.26) \times 0.09 \times 7 \times 640 \times 225$$

$$= 2{,}924{,}304{,}768 \text{ ft.}^3$$

$$= 2.9 \text{ bcf}$$

Zone 3:

$$G_p = 43{,}560 \times (1 - 0.24) \times 0.065 \times 5 \times 640 \times 225$$

$$= 1.55 \text{ bcf}$$

Zone 8L (lower): $G_p = 1.45$ bcf

Zone 8U (upper): $G_p = 2.98$ bcf

For the lower Cunningham, the total gas reserves are 4.45 bcf; for the Morrow, 4.43 bcf.

Making the same analysis that we made in Chapter 7 for a 78% net revenue interest (NRI) and a sale price of $1.25/Mcfg produces the following table:

	Best Case	Most Likely	Worst Case
Well cost	($1,200,000)	($1,200,000)	($1,200,000)
Net income LC	3,900,000	2,900,000	2,300,000
Net income Morrow	3,880,000	1,725,000	—
Add'l $ to complete Morrow	(150,000)	(150,000)	(150,000)
Profit	$6,430,000	$3,275,000	$950,000
Return on investment	4.76	2.42	0.70

This is a marginal well, according to the guidelines for Uncle Howard's company. They like to have the prospect of a 4:1 return before completing, but only in the best case will the well have that good a return.

In real life, this well was completed in the lower Cunningham only. It looks as if it will make about 2.5 bcf. In addition, the gas price has dropped to $1.10/Mcf. (Average price for the life of the well will be about $1.15.) If the Morrow is not completed, the ROI will be 1.69. Obviously the Morrow can be completed fairly cheaply, and this will be done in the future.

GULF COAST EXAMPLE

Now let's look at a suite of logs from the Gulf Coast region.

We have another gas well. The formation is an unconsolidated, high-porosity sandstone. The zone of interest is the upper Hull sand, a prolific producer in the area. We want to estimate reserves.

We have a combination of logs: Induction Electrolog/BHC Acoustilog (Western Atlas product names; formerly Dresser Atlas) with SP and caliper curve (Plate 6) and compensated neutron-density porosity log with GR and caliper curves (Plate 7). The neutron and density logs were run on sandstone matrix (noted on the heading under logging data) because we are in an area that is predominantly sandstones.

Look carefully at the two logs and notice where the top of the formation is marked. There is an 8-ft. difference between the two logs; one was not recorded properly. Mistakes like these must be recognized; otherwise, even more serious mistakes may be made, such as using porosity data with resistivity data from different parts of the formation.

The first thing to do is determine R_w. Look at zone A at 9860–9866 on the Induction Electrolog. This zone is obviously wet because the resistivity is very low (0.4 ohms) and the SP has the maximum deflection seen on the logs. Find the equivalent point on the porosity logs at 9868–9874. Read the porosity; both the neutron and the density read the same, which is appropriate in a clean, wet sand. Porosity is 28%. So

$$R_o = F_R \times R_w \text{ or}$$

$$R_w = R_o / F_R \text{ but}$$

$$F_R = 0.81/\Phi^2 \text{ for unconsolidated formations, so}$$

$$R_w = R_o \times (\Phi^2/0.81)$$

$$R_w = 0.40 \times (0.28^2/0.81) = 0.039 \text{ at BHT}$$

You can check R_w by calculating it from the SP. In Plate 5, the shale base line and the maximum deflection at a sand (SSP) are already drawn for you. The shale base line is drawn through a uniform shale bed and extends above and below the zone of interest. The SSP is drawn through the largest deflection that is noted opposite a sand. In this case, the maximum deflection of –5.7 divisions is at Zone A.

To calculate R_w from the SP you need the SP deflection in millivolts (–5.7 div x –10 mV/div = –57 mV); R_{mf} at measured temperature (0.38 at 85 °F); and bottom-hole temperature (188 °F from the heading). First you must convert R_{mf} at 85 °F to BHT:

$$R_{mf} \text{ @ BHT} = 0.38 \times (85 + 7)/(188 + 7) = 0.18$$

Next, find R_{mf} on Figure 8–1 and convert it to R_{mfe} (0.14). With R_{mfe} and SP = –57 mV, determine R_{mfe}/R_{we} = 4.8 from Figure 8–2. By continuing through the chart, you'll come out with R_{we} = 0.027 ohms.

Go back to Figure 8–1, this time with R_{we} on the left edge, and find R_w = 0.037. This value compares well with R_w by the R_o method. Compromise and use 0.038 for R_w.

Now set up another log analysis table (Table 8–2). The listings will be slightly different because there is an additional curve, sonic traveltime (t). Since the porosity is high, the invasion will be shallow and the induction resistivity will be very close to R_t. The sonic porosity will be too high because of gas and shale effects, but it gives a good quick look using the R_{wa} curve in track 1 of Plate 6.

FIELD Evans Ranch					COUNTY							STATE		
DEPTH	R_D	T	Φ_s	GR	V_{sh}	Φ_D	Φ_N	Φ_{xp}	Φ_e	Φ-ft.	S_w	Hc-ft.	BVW	REMARKS
9856	1.0	95	30	2.4	0.05	24	30	27	25.5	–	69	–	.175	Wet
52	0.3	96	30.5	3.3	0.29	25	25	25	16.5	0.66	100	0	.25	Wet
48	.45	94	29	2.6	0.11	26	30	28	25	1.00	100	0	.25	Wet
44	2.0	96	30.5	2.2	0	30	23	27	27	1.08	46	0.58	.15	Wet
40	8.5	96	30.5	2.2	0	32.5	12	24.5	24.5	0.98	25	0.74	.06	Gas
36	10.0	84	21.5	2.2	0	34	13.5	26	26	1.04	21	0.82	.055	Gas
32	20.0	92	27.5	2.4	0.05	32.5	12	24.5	24.5	0.98	25	0.74	.06	Gas
28	12.0	84	21.5	2.7	0.13	25.5	14.5	21	17.5	0.70	22	0.55	.039	Gas
24	32.0	89	25	2.2	0	32	10.5	23.5	23.5	0.94	13	0.82	.030	Gas
20	21.0	70	12	2.3	0.03	13.5	9	12	11	0.44	31	0.30	.034	Gas
16	15.0	82	20	2.5	0.08	11	11	11	9	0.36	36	0.23	.032	Gas
12	12.0	94	29	2.7	0.13	21	12	17.5	14	0.56	26	0.41	.037	Gas
08	2.6	103	36	3.0	0.21	7	12	9.5	3.5	–	–	–	–	Tight
04	6.0	60	3	2.7	0.13	0	9	3.6	1	–	–	–	–	Tight
Shale	0.9	96	30.5	6.0	1.0	23	34	27	–	–	–	–	–	Shale Pt.

For Φ_{xp} use: $\Phi_D = (1.5\ \Phi_D + \Phi_N)/2.5$

Table 8–2. *Log Data for Shaly Sand Example*

R_{wa} is the apparent formation water resistivity. Remember from the Archie equation that $R_o = F_R \times R_w$ or $R_w = R_o / F_R$. If you assume that every zone is wet, you can calculate a range of values for R_{wa}. The lowest values will correspond to the actual wet zones, and higher values will correspond to oil or gas zones. By plotting these values on the log, the high apparent R_w values will stand out, showing you where to make your detailed calculations. You can easily make this calculation because the induction log was run in combination with a sonic log. The sonic traveltime is converted to Φ and then to F_R. The induction resistivity is divided by F_R; the output is R_{wa} plotted in track 1. The scale for R_w on Plate 6 is 0–1.0. Note how it is around 0.03 in the wet sands and then how it jumps to very high values through the zone from 9844 to 9806 ft. This indicates the zone is probably productive.

Use the GR log to calculate V_{sh}. GR_{sh} = 6.0 divisions and GR_{cl} = 2.2 divisions (see the lines drawn on Plate 7).

For the detailed analysis, use the crossplot porosity from the neutron density log. Although there are charts that help us find Φ_{xp}, it is just as accurate and much quicker to use the equation

$$\Phi_{xp} = ((1.5 \times \Phi_D) + \Phi_N)/2.5$$

This relationship should be used in gas zones with reservoir pressures $<$ 5000 psi. The density porosity is weighted because it is less affected by the gas than is the neutron porosity.

In high-pressure gas zones the following equation should be used:

$$\Phi_{xp} = (\Phi_D + \Phi_N)/2$$

Φ_e is calculated from the relationship

$$\Phi_e = \Phi_{xp} - (Vsh \times \Phi_{shxp})$$

The next column in Table 8–2 contains a new parameter: porosity-feet. This parameter is the effective porosity multiplied by the depth interval between calculations. Here you are making a calculation every 4 ft. (in reality, you'd make one every 2 ft.), so multiply $\Phi_e \times 4$. To find the total porosity-feet for the zone, add the Φ-ft. column (8.66). If you want the average porosity for the zone, divide by the zone thickness (44 ft.) so that Φavg = 8.66/44 = 0.197 = 19.7%.

Next, calculate the water saturation for each point. The presence of shale in the formation reduces the true resistivity. If you were to use the Archie equation for S_w, you would get values that are too high and you might miss a productive zone. Many different methods have been developed to calculate shaly sand water saturations; the method you'll use here was developed by Simandoux and is relatively straightforward:

$$S_w = \frac{c \times R_w}{\Phi^2}\left[\sqrt{\frac{5\Phi^2}{R_w \times R_t} + \left(\frac{V_{sh}}{R_{sh}}\right)^2} - \left(\frac{V_{sh}}{R_{sh}}\right)\right]$$

where:

c = 0.4 for sands and 0.45 for carbonates

Note that when $V_{sh} = 0$, you can use the simpler Archie equation. Although the Simandoux equation is cumbersome in hand calculations, even a simple programmable calculator can handle it easily.

Now calculate S_w using R_{ID} for R_t, $R_w = 0.038$, and $R_{sh} = 0.9$ with the values for V_{sh} and Φ_e from Table 8–2. List your results in the S_w column of the table.

Now we come to another new parameter: hydrocarbon-feet (hc-ft.). This is porosity-feet times $(1 - S_w)$. (Remember that the hydrocarbon saturation $S_h = 1 - S_w$). From hydrocarbon-feet you can find the average water saturation, and you can use the measurement very conveniently when you make your reserve calculation.

The final column in Table 8–2 is bulk volume water (BVW). On inspecting the column, you see that there is a gas/water contact at 9844–9840 ft. Call 9840 the bottom of the productive zone. The top is 9812; above that, the zone is tight.

Now let's perform a reserve calculation. This is a water-drive reservoir, and the recovery factor is estimated to be 80%. Use $1/B_g = 175$ (the gas expansion factor),

$$\text{Reserves} = 43{,}560 \times [h \times \Phi \times (1 - S_w)] \times A \times 1/B_g \times RF$$

The term in the brackets is the hydrocarbon-feet; you can obtain this from Table 8–2. The total hydrocarbon-feet value from 9840–9812 is 5.21 – 0.58 = 4.63:

$$\text{Reserves} = 43{,}560 \times 4.63 \times 160 \times 175 \times 0.8$$
$$= 4.52 \text{ bcf}$$

In a nutshell, this is how you would work through a detailed log interpretation. Today, however, computers help us generate and interpret logs. To learn more, read Chapter 9.

9

Computer-Generated Interpretations

The amount of information available from well logs today is truly amazing, especially when you consider that 20 or 30 years ago a sophisticated interpretation was an S_w calculation made by hand every 2 ft. New radiation detectors, downhole microprocessors (minicomputers), and especially the truck-mounted (onboard) computer have had the most profound impact on the logging industry since the first log was run in Pechelbronn, France, in 1927.

The new sensors and microprocessors have made a multitude of measurements possible: gamma-ray spectroscopy, induced spectroscopy, bulk density with photoelectric effect, sonic waveform recording, and stratigraphic dip measurements, to name a few. Most of these measurements were either unknown or impossible to make except in laboratories only a few years ago.

Powerful onboard computers have made all of these measurements possible. The computer processes the mass of information that is transmitted to the surface by the downhole telemetry system, stores it on magnetic tape, corrects for the environment, merges the data from several different tools, makes complex calculations, and prints out the information in various log formats while simultaneously keeping the logs on depth and warning the logging engineer if a tool malfunctions. The wellsite computers are able to transform all this data into a variety of presentations from conventional logs to pictorial representations of the wellbore and formations—often in color.

It is even possible to transmit the data from the wellsite via satellite to a computing center or oil company office anywhere in the world.

Without a doubt, the computer is a wonderful labor-saving device. It performs calculations and makes data presentations that would be impractical by hand. The computer's strengths are its speed and ability to make repetitious calculations; one of its weaknesses is information overload. The tremendous amount of data, often in new and unfamiliar forms, can be overwhelming. Another weakness of the computer is that it still does only what it is told. It cannot think (at least not yet) and cannot make judgments or true interpretations. We still need humans for that.

WELLSITE COMPUTER LOGS

Wellsite computer logs are often called quick-look logs. These logs are generated at the wellsite after (or sometimes during) the data collection phase—after the tools have been run in the well and the data have been collected. The logs may be made from calculations using previously collected data from several different devices, or they may be just a different presentation of the data from a single log. For example, Figure 9–1 is an alternative presentation of the caliper information in a form more useful to the drilling engineer. Here, the volume of cement needed has been calculated from the caliper information for a specific casing size. The geologist or engineer uses wellsite computer logs like this one to make a rapid evaluation of the formations or the well condition. All of the major logging companies and many of the smaller ones have onsite computing capabilities and furnish similar onsite products.

Figure 9–2 is the wellsite computer-generated interpretation log for the Sargeant 1–5. This is an excellent example of both the strengths and weaknesses of a computer-generated log. In this case, an R_w of 0.09 was used throughout the lower Cunningham and the Morrow. You know, both from offset production and from previous calculations on the logs, that R_w is closer to 0.12 for the lower Cunningham and 0.2 for the Morrow. Using 0.09 instead of the correct R_w will give water saturations that are 15% too low for the lower Cunningham and 33% too low for the Morrow. This variation is large enough to cause a major mistake, such as running pipe on a wet well or abandoning a productive well.

The major precaution to take when looking at someone else's interpretation—even a computer's—is to check all of the assumptions, inputs, and

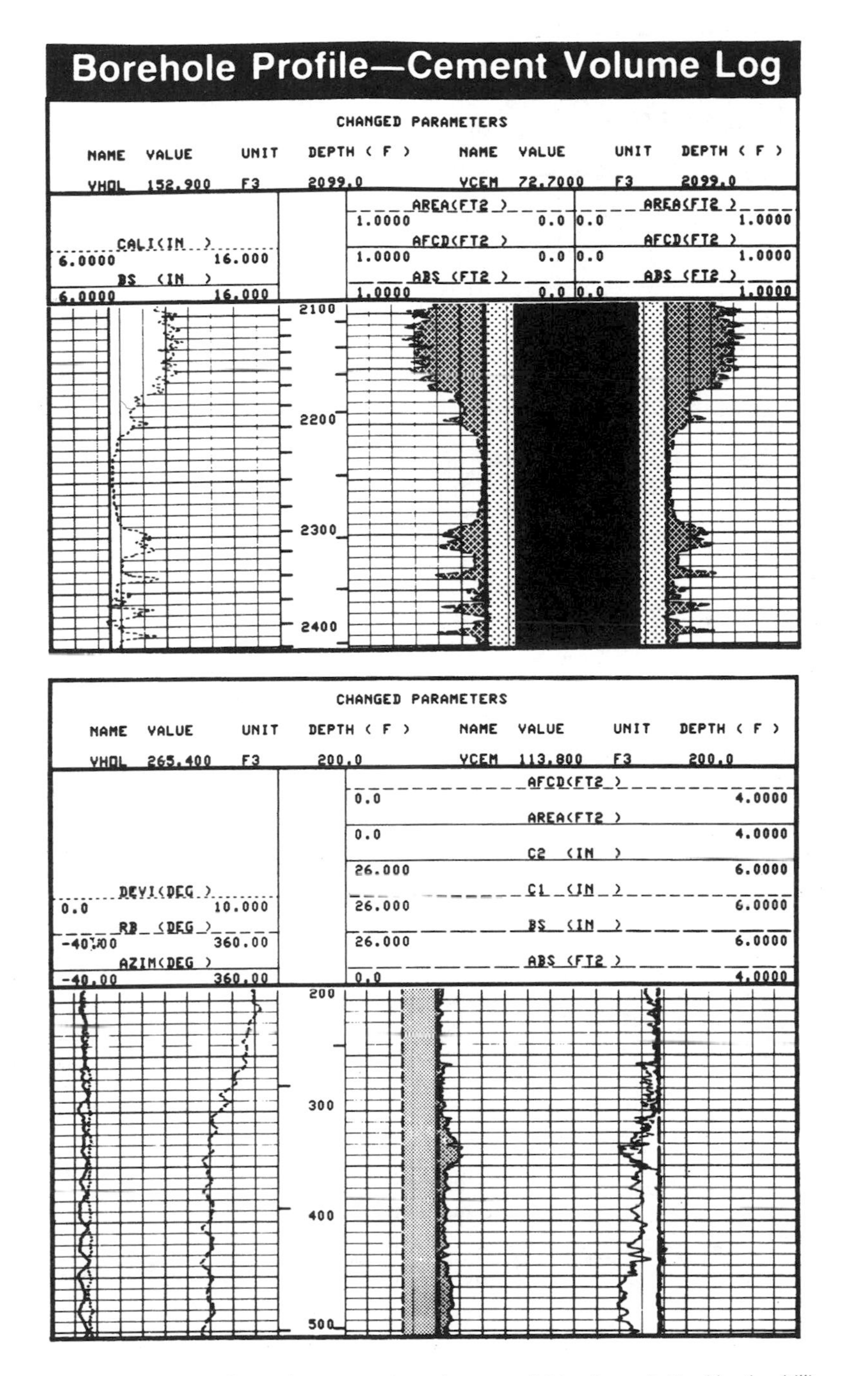

Fig. 9–1. *Borehole profile and cement volume (courtesy Schlumberger). Used by the drilling engineer to determine how much cement will be needed to cement the casing.*

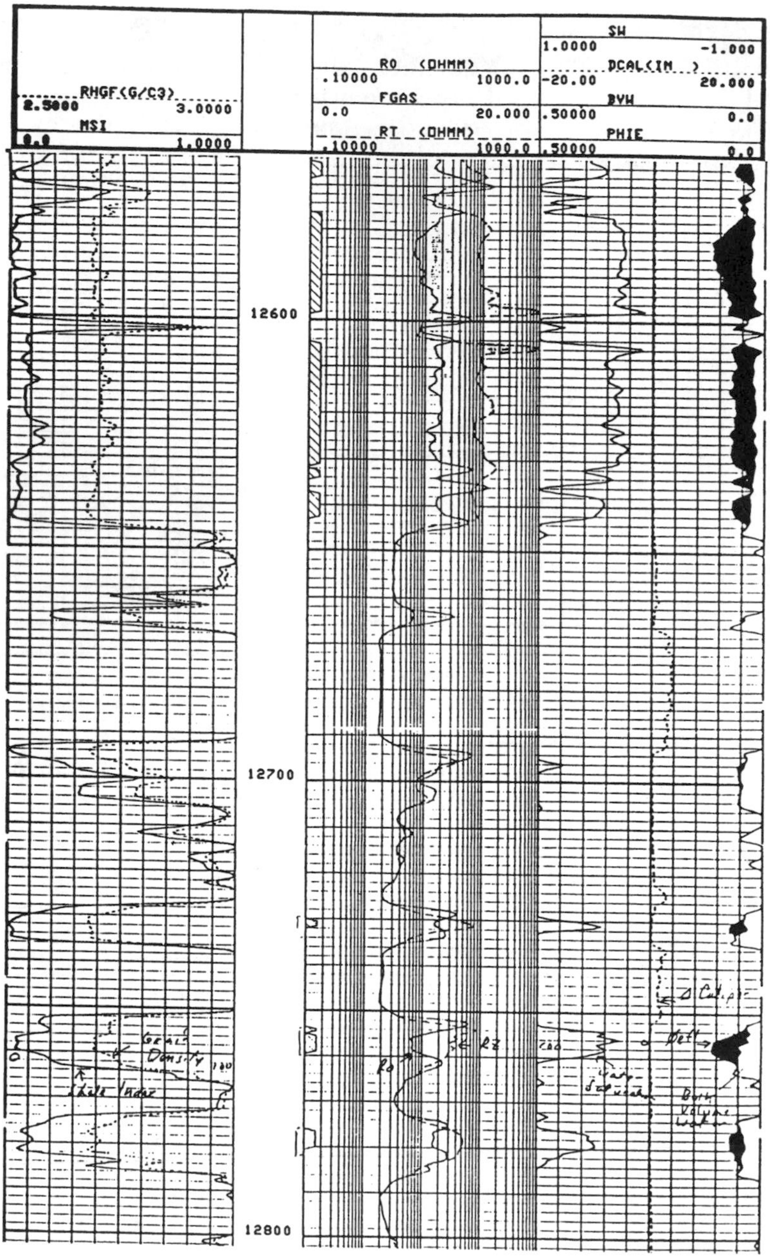

Fig. 9–2. *Wellsite computer-generated log for the Sargeant 1–5. Note that the water saturations are lower than those calculated in Chapter 8. See the text for the explanation.*

methods used. Why was a particular R_w chosen? Does it agree with what was used in other wells in the same zone? What was used for the shale values? Were log values corrected for environmental effects? These questions are important because they affect the corrections made to the log readings in the zones of interest. Then we can ask whether the well was properly zoned. In other words, were the necessary parameters changed as the interpretation was made in a different formation?

The most common onsite computer-interpreted log uses measurements of deep resistivity, neutron density porosity, GR, SP, and caliper curves to solve for water saturation, crossplot porosity, and lithology. The log values are first corrected for environmental effects (temperature, borehole size and salinity, and mud weight) in a preliminary interpretation pass. On this pass (Fig. 9–3), the corrected porosities from the neutron and density logs are crossplotted. The logging engineer then picks the various parameters needed (R_w, GR_{sh}, SP_{sh}, Φ_{Nsh}, etc.) and makes the interpretation pass.

Figure 9–4 is the interpretation pass for the well in Figure 9–3. Notice, on both figures, the photoelectric factor (PEF) curve in track 3. The PEF curve indicates lithology, shown in the depth track as limestone (no coding), sandstone (coarse dot pattern), and dolomite or shale (fine dot pattern). Also note on Figure 9–4 that track 1 includes a dashed coding for the shales. The area between Φ_e and BVW in track 3 is coded black to indicate hydrocarbon. The log shows promise at the following zones:

Zone 1 at 6354–6360 with Φ = 12% and S_w = 30%
Zone 2 at 6338–6346 with Φ = 13% and S_w = 15%
Zone 3 at 6230–6240 with Φ = 12% and S_w = 35%

These zones will need further investigation.

Obviously, the major strength of computer-generated interpretation logs is that calculations are made continuously; all we have to do is read values of Φ_e, S_w, and BVW directly from the log (after we have checked the assumed values). The computer helpfully shades the area between BVW and Φ_e so that the amount of hydrocarbon stands out, or it indicates different rock types by changing the shading or dot pattern between curves.

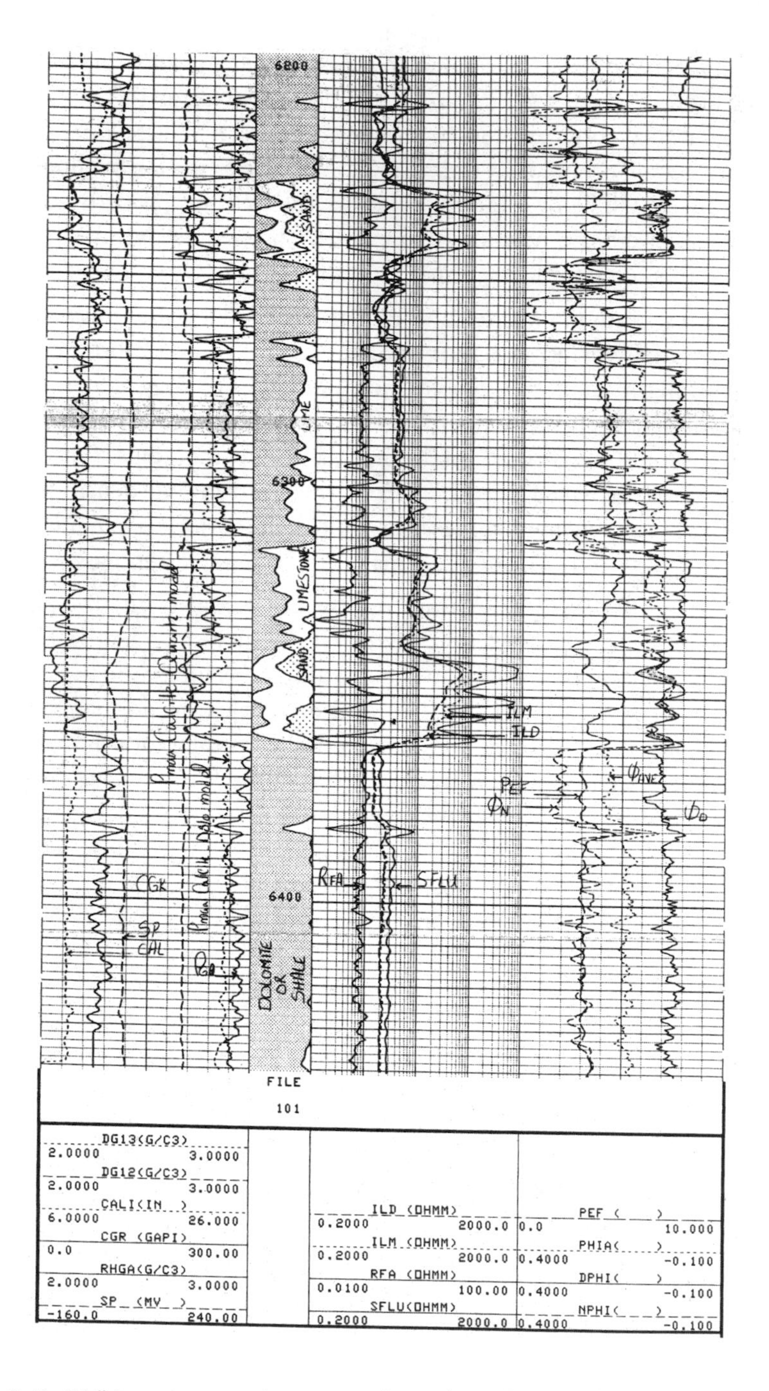

Fig. 9–3. *Wellsite preinterpretation pass, used to make environmental corrections and to pick the parameters used in the final interpretation pass.*

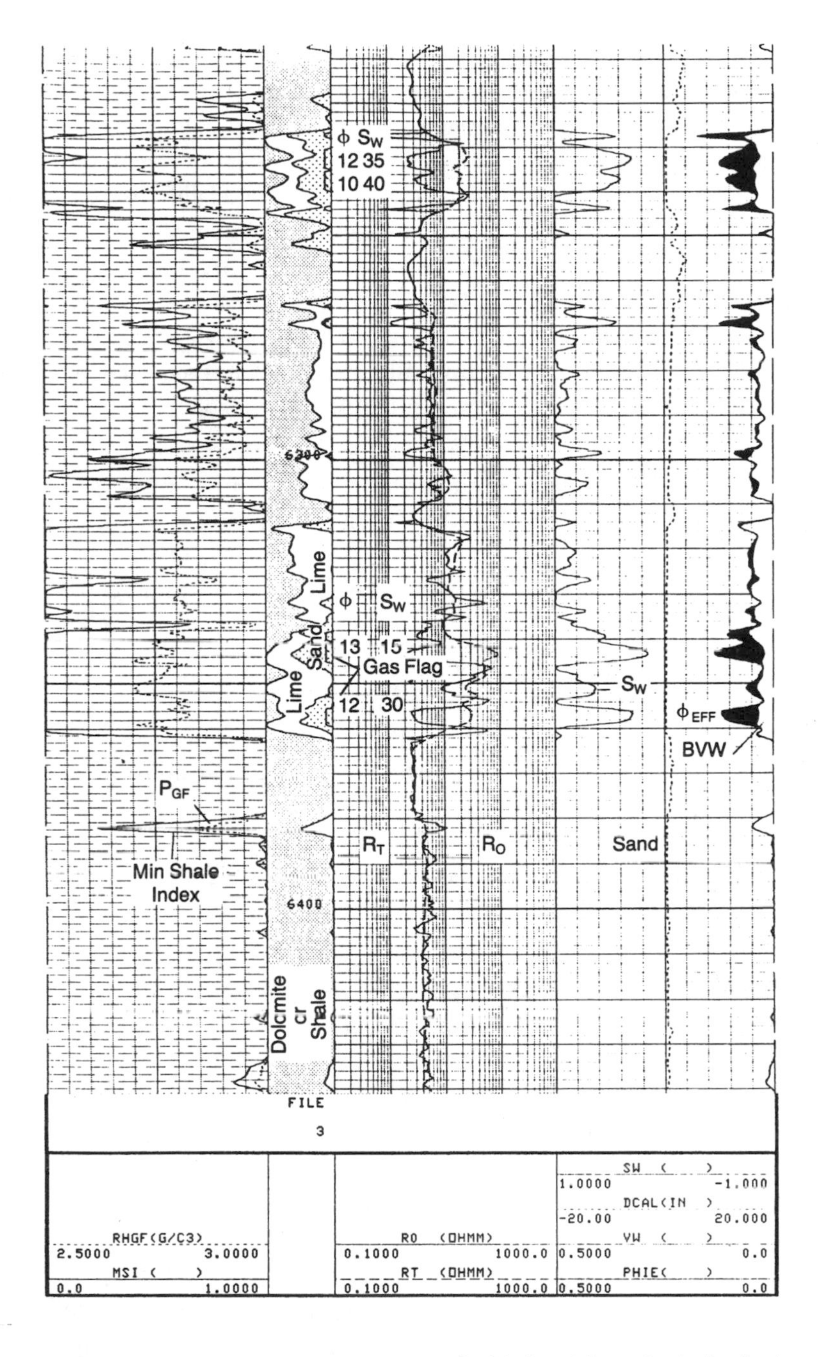

Fig. 9–4. *Wellsite computer-interpreted log. Note the lithology information in the depth track.*

COMPUTING CENTER LOGS

Computer logs are generated either at the wellsite or in computing centers. The computing centers are staffed with experienced log analysts and have powerful interpretation programs. They often process the more complex logs such as Schlumberger's Formation Micro Imager (FMI) or Halliburton's Electrical Micro Imaging Service (EMI), where an image of the wellbore is interpreted by an experienced engineer looking for dips or fractures—techniques that require human expertise. The computing centers also make a more sophisticated interpretation than is possible in the field. One technique simultaneously considers the various input or information from all sources (not just log data) and calculates the best-fit results using as few assumptions as possible.

Compare Figure 9–4, a wellsite computer-interpreted log, with Figure 9–5, a log generated and interpreted at a computing center—same well, same interval. Immediately we notice the change in format. The depth track is on the left edge of the log. GR and apparent grain density curves ρ_{ga} (determined by solving the density porosity equation for grain density with Φ_{xp} and ρ_b as inputs) are in track 1, as is a permeability indicator.

Track 2 in the center of the log is divided into two subtracks: track 2A shows water saturation (and residual hydrocarbon volume if a tool such as a microlaterolog was run), and track 2B shows permeability (k) and BVW. The caliper curve is presented with the bit size at 0 divisions on track 2B. Track 3 shows the bulk volume analysis of the entire formation. In this case, the bulk volume is divided into dry clay, bound water (water associated with the dry clay), silt, matrix, and effective porosity. The porosity is further divided into water and hydrocarbon. The presentation varies depending on the logs that were run.

Water saturations are slightly higher and porosities are lower on the computing center log. Zone 1 now has a maximum porosity of 8% with a water saturation of about 40%. The zone has shrunk to about 2 ft. of pay, which is not very promising. Zone 2 has a maximum porosity of 9% with a water saturation of 20%. We still have about 4 ft. of pay. Zone 3 has about 2 ft. of 9% porosity with a water saturation of 30%.

The computing center logs are considered to be more accurate interpretations than field-generated computer logs, mainly because the powerful computers and programs used in the computing centers use more information and apply more complex corrections than at the wellsite. In addition, a log analysis specialist makes the interpretation, often with input from the oil company log analyst or geologist.

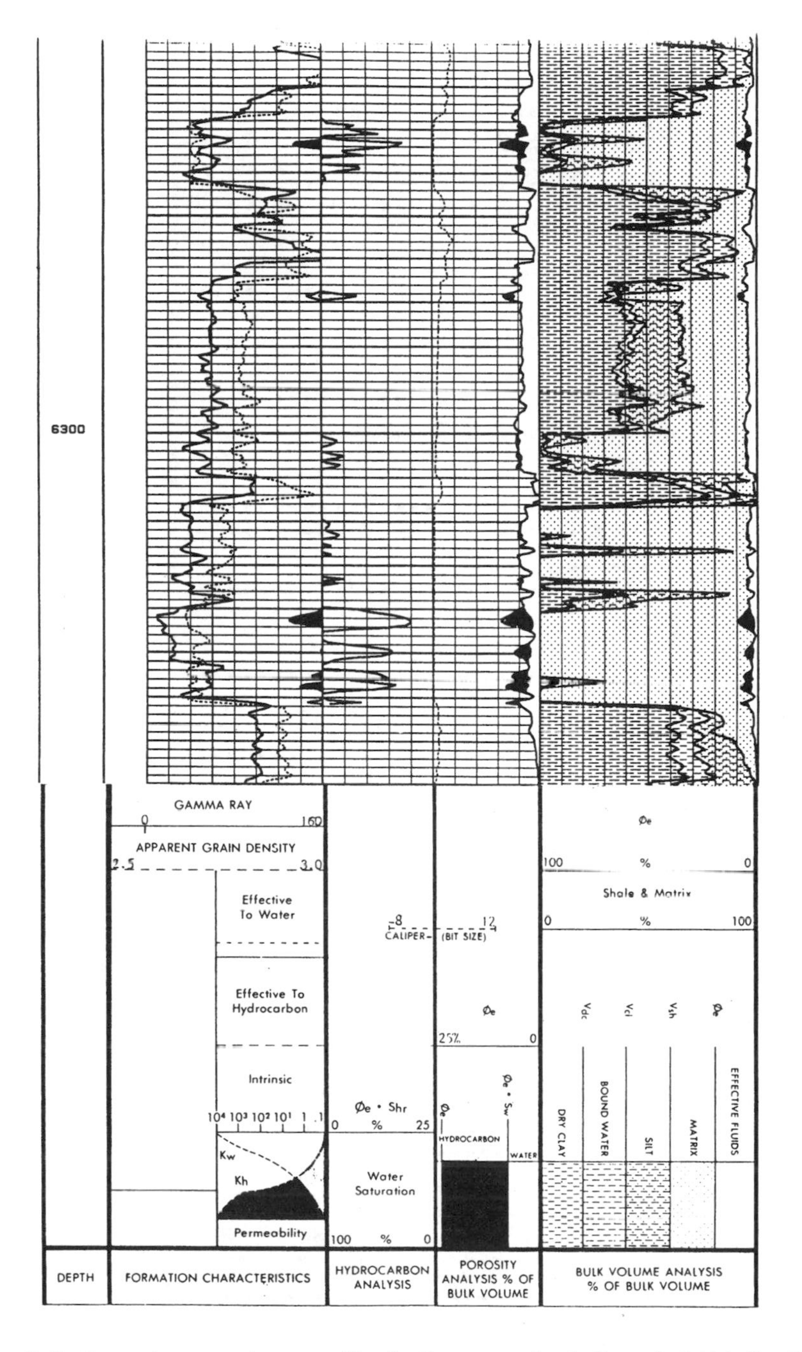

Fig. 9–5. *Computing center interpreted log for the same well as in Figure 9–4. Note the different format with the depth track on the left. This is a convention adopted by logging companies to distinguish computing center logs from wellsite logs.*

COMPANY : SALES EXAMPLE WELL : PAGE 5

DEPTH FT	WAT. SAT. PU	POR. EFF. PU	POR. 2ND PU	CUM. EFF. POR. FT	CUM. HC-FT FT	PERM. INTRIN MD	PERM. GAS MD	PERM. OIL MD	PERM. WATER MD	CUM. PERM. GAS FT	CUM. PERM. OIL FT	CUM. PERM. WATER FT	CLEAN GRAIN DENS G/C3	VOL SH PU
6268	85	8.		4.1	.6	.2	0	0	0	6.	0	0	2.70	37
6269	79	8.		4.0	.5	.2	.1	0	0	6.	0	0	2.70	36
6300	100	5.		2.7	.5	0	0	0	0	6.	0	0	2.80	12
6302	87	5.		2.6	.5	0	0	0	0	6.	0	0	2.84	15
6334	66	5.		1.5	.5	0	0	0	0	6.	0	0	2.79	3
6340	67	5.		1.3	.5	0	0	0	0	6.	0	0	2.71	16
6341	42	7.		1.2	.4	.3	.3	0	0	6.	0	0	2.69	0
6342	23	9.		1.2	.4	2.5	2.5	0	0	5.	0	0	2.69	0
6343	20	8.		1.1	.3	1.9	1.9	0	0	3.	0	0	2.69	0
6344	39	5.		1.0	.3	0	0	0	0	1.	0	0	2.70	0
6349	41	5.		.9	.2	0	0	0	0	1.	0	0	2.71	0
6350	38	6.		.8	.2	.1	.1	0	0	1.	0	0	2.71	0
6355	45	7.		.6	.1	.3	.3	0	0	1.	0	0	2.69	0
6357	34	8.		.5	.1	.6	.6	0	0	1.	0	0	2.69	13
6358	52	6.		.4	0	.2	.1	0	0	0	0	0	2.70	0
6361	65	8.		.3	0	.5	.2	0	0	0	0	0	2.69	24

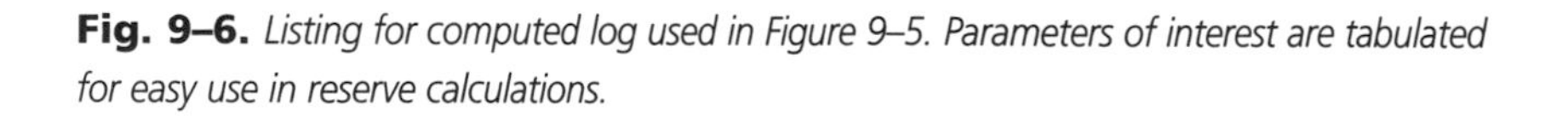

Fig. 9–6. *Listing for computed log used in Figure 9–5. Parameters of interest are tabulated for easy use in reserve calculations.*

Also included with the more advanced interpretation is a listing (Fig. 9–6) of some of the important outputs (S_w, Φ_e, permeability, grain density, and V_{sh}) as well as some additional calculations (cumulative porosity-feet and hydrocarbon-feet) that are useful for reservoir and hydrocarbon-in-place calculations. Look at the zone at 6340–6344 ft. on the computed log; then find the zone in the listing. S_w varies from 67 to 20% and Φ_e from 5% to 9%. The column for cumulative porosity-feet sums up (integrates) the porosity versus depth. To find the number of porosity-feet for this zone, subtract 0.9, the cumulative porosity-feet at the zone just below 6344, from 1.3, the total porosity-feet for the well from total depth to 6340. The result is 0.4 porosity-feet. Out of 4 ft. of formation, 0.4 ft. of space is available to hold fluids. This value can be used to draw net isopach maps (net porosity maps).

The next column in the listing is cumulative hydrocarbon-feet, which we get by multiplying $(1 - S_w) \times \Phi_e$ for each foot of formation. To arrive at this number, subtract the value for hydrocarbon-feet at the last depth below the zone of interest (0.2 hc-ft. at 6349 ft.) from the number of hydrocarbon feet at the top of our zone (0.5 hc-ft.). The result is the term $(1 - S_w) \times \Phi_e \times h$ in the reserves calculation. To calculate the reserves for this 4-ft. zone,

$$G_p = 43{,}560 \times (1 - S_w) \times \Phi_e \times h \times A \times (1/B_g)$$

But

$$(1 - S_w) \times (\Phi_e \times h = \text{cumulative hydrocarbon-feet for the zone.}$$

So

$$\begin{aligned} G_p &= 43{,}560 \times (0.5 - 0.2) \times 160 \text{ acres} \times 125 \\ &= 0.262 \text{ bcf} \end{aligned}$$

So far, we have relied on traditional measurements—resistivity and porosity—to determine water saturation. This requires a resistivity profile, two or three independent porosity measurements, and an estimate of the formation water salinity. We can use resistivity to calculate water saturation because of the large contrast in resistivity between a formation saturated with oil and one saturated with saltwater (distilled water has an infinite resistivity just like oil or gas). The normal case is for formations to contain saline water of moderate to high concentration. When this is not the case and the salinity is low, the resultant resistivity is high and there is little contrast in the readings between oil-bearing and water-bearing formations.

Another problem occurs when formation water salinity is not constant. This can occur in a waterflood when the injected water differs from the original formation water. In other instances, formation water salinity may vary greatly over a short interval, as we saw in the R_w change between the Cunningham and the Morrow.

Obviously, a measurement of water saturation that is not affected by water salinity is needed. With new tools, we have a choice of methods to use on those formations that, in the past, refused to yield to traditional interpretation methods. The first method we'll look at is the dielectric constant technique.

DIELECTRIC CONSTANT LOGS

Table 9–1 lists laboratory-determined values of the dielectric permittivity ε of some common formations, minerals, and reservoir fluids. Most of the dielectric values are low except for water. Although the value for water varies somewhat with salinity, it is still nearly 25 times greater than oil and 50 times greater than gas. This is a good contrast in value that is nearly independent of salinity. Since the dielectric constant for the formations is low and almost constant for all of the formations except shale, the variations in dielectric permittivity measurements are mainly a function of water-filled porosity.

Name	Relative Dielectric Constant	Propagation Time t_{pl} (ns/m)
Sandstone	4.65	7.2
Dolomite	6.8	8.7
Limestone	7.5–9.2	9.1–10.2
Anhydrite	6.35	8.4
Halite	5.6–6.35	7.9–8.4
Gypsum	4.16	6.8
Dry colloids	5.76	8
Shale	5–25	7.45–16.6
Oil	2–2.4	4.7–5.2
Gas	1	3.3
Water	56–80	25–30
Fresh water	78.3	29.5

Table 9–1. *Relative dielectric constants and propagation times*

By measuring the attenuation and phase shift of an electromagnetic wave and knowing the angular velocity and magnetic permeability from the tool design, we can determine the dielectric constant and conductivity of the formation.

The basic equations describing electromagnetic propagation are

$$\gamma = \alpha + j\beta$$
$$\omega^2 \mu \varepsilon = \beta^2 - \alpha^2$$
$$\omega \mu C = 2 \alpha \beta$$
$$t_{pl} = \beta / \omega$$

where:

γ = electromagnetic wave propagation
α = attenuation of the wave
β = phase shift
ω = angular velocity
μ = magnetic permeability
ε = dielectric constant
C = conductivity
t_{pl} = propagation time of the wave

The type of tool designed to measure the dielectric constant has microwave transmitters and receivers mounted in a pad pressed against the wall of the wellbore. The frequencies used are in the range of 1.0 gigahertz. The wave is sent out from the two transmitters, and the attenuation and phase shift are measured at the antennas (see Fig. 9–7). These two measurements, attenuation and phase shift, are the raw data recorded on the log. The pad-mounted tool reads mainly in the flushed zone.

Another dielectric measuring tool design uses a lower-frequency electromagnetic wave. The transmitter and receiver antennas are all mounted on a centralized mandrel rather than a pad. The tool uses several receivers spaced along the mandrel so that an invasion profile can be determined. Both tools usually include both a microlog and a caliper.

Interpretation

Variations in the dielectric constant are primarily due to changes in the amount of water present in the formation. This water can be mud filtrate in the flushed zone, original formation water, a mixture of the two, or bound water in the shales. Because of the shallow depth of measurement, we assume that usually the tool is reading the flushed zone.

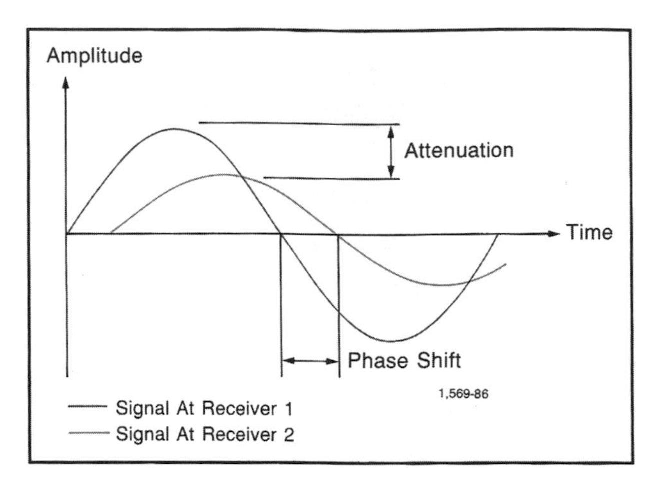

Fig. 9–7. *Electromagnetic propagation signals (courtesy Schlumberger). The attenuation and phase shift between receivers is the basic measurement of the EPT.*

The propagation of the electromagnetic wave through the formation can be calculated with a weight average equation similar to the one used for the density log:

$$\gamma = \Phi\, \gamma_f + (1 - \phi)\, \gamma_{ma}$$

where:

γ = resultant electromagnetic propagation wave in the formation
γ_f = electromagnetic wave propagation in the fluid within the pore space
γ_{ma} = electromagnetic wave propagation in the matrix
Φ = porosity

Several different interpretation schemes are used for the dielectric constant measurement. All of them are heavily computer dependent. Corrections for different phenomena such as geometrical spreading and scattering must be made to the tool readings. All of the interpretation methods yield similar results. An apparent porosity is calculated that is mainly a measurement of the water-filled porosity. By comparing this dielectric constant porosity, or Φ_{dc}, to the total porosity as measured by a neutron density crossplot, for example, a quick-look estimate of the flushed-zone water saturation can be made. The value of Φ_{dc} will be the same as the crossplot porosity Φ_{xp} in water-bearing zones ($S_w = 100\%$). In hydrocarbon-bearing zones crossplot porosity will be greater than dielectric constant porosity.

In Figure 9–8 an electromagnetic propagation tool (EPT) log is combined with GR, dual laterolog, neutron, and density logs. In this South American well, the productive zone is located in a freshwater formation. The resistivity logs are of little help in locating the oil/water contact. However, by comparing the EPT porosity to the neutron density crossplot porosity, the contact is readily apparent at 6850 ft. The EPT reads only the water-filled porosity and not the porosity that contains oil. Below the contact at 6850, the water saturation is 100% and the EPT and the crossplot porosity are the same.

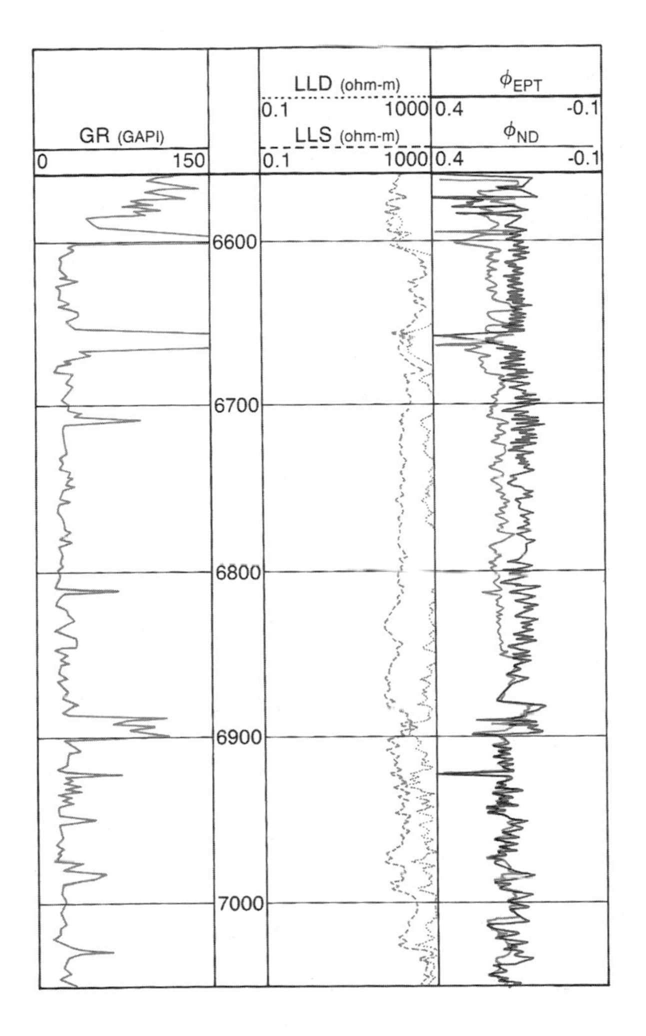

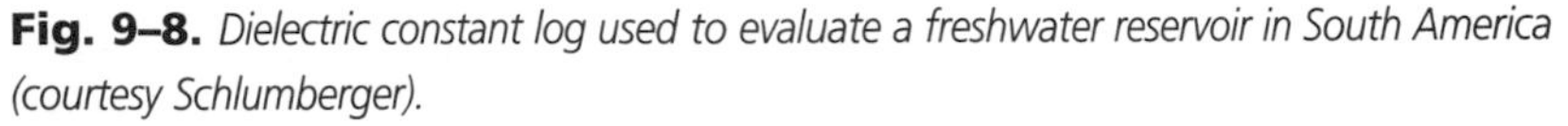
Fig. 9–8. *Dielectric constant log used to evaluate a freshwater reservoir in South America (courtesy Schlumberger).*

In the advanced computer-generated interpretations, the dielectric-constant porosity is used to calculate the amount of moved hydrocarbons, to determine accurate flushed zone water saturations S_{xo}, and to give a better estimation of lithology and clay type.

MAGNETIC RESONANCE LOGS

Most of us have heard of MRIs, or magnetic resonance imaging, in relation to the medical field. The same technology but in a much smaller form is also available to the petroleum industry. The technique was initially called nuclear magnetic resonance, or NMR. The terms NMR and MRI are often used interchangeably. NMR was first tried in 1960 but with limited success. The attraction of the technique was its promise to measure the free fluid index (FFI), which is the percent of fluid that is producible. Over the years, NMR logging equipment has been refined repeatedly. Today we have a technique that determines fluid-filled porosity and, sometimes, total porosity independently of the lithology. Several different tool designs are in use, depending on the logging company. All of them, however, use the same physics of measurement.

Nuclear magnetic resonance, as used in well logging, refers to the response of the formations to an imposed magnetic field. As we know from our high school physics courses, the nuclei of atoms (mainly the protons) are affected by magnetic fields. The nuclei in a magnetic field behave essentially like gyroscopes in a gravitational field. A strong magnet will cause the protons to align with the magnet's field. If another magnetic field is applied at right angles to the first field and then removed, the protons will try to realign with the original field by the mechanism of *precession*. While they precess, the protons emit a measurable signal. Hydrogen atoms produce a fairly large signal, while other formation elements generally have small signals. Fortunately, hydrogen occurs in abundance in formation fluids. By tuning the logging tool to the resonant frequency of hydrogen, we are able to make a measurement proportional to the porosity of the formation, independent of the lithology.

The tools use a strong permanent magnet that aligns the hydrogen protons in a magnetic field referred to as B_0. A transmitter emits a radio frequency magnetic pulse that causes the hydrogen protons to tip 90°. After a certain time, another magnetic energy pulse is applied at 180° to the first pulse. This causes the protons to spin, or precess. Several hundred pulses, each 180° out of phase to the preceding pulse, are made at equal intervals. The precession, or *spin echo*, of the protons generates a signal that is measured by the tool circuitry. The initial amplitude of the signal is a measure of the volume of formation fluids. The decay rate (in amplitude) of the signal versus time is called the *transverse relaxation time*, T_2. The faster the decay rate (shorter time), the lower the permeability.

Figure 9–9 shows the T_2 distribution for a sandstone. The portion of the signal due to clay-bound water appears first. This is followed by the capillary bound (irreducible) water and then by the producible (movable) fluids. The cutoffs, such as CMR free fluid porosity represented by the vertical dashed line in the figure, are somewhat arbitrary and vary with the formations encountered. Not all tools are able to measure the clay-bound water fraction. The clay-bound portion of the signal appears very early in the measurement cycle and is very small.

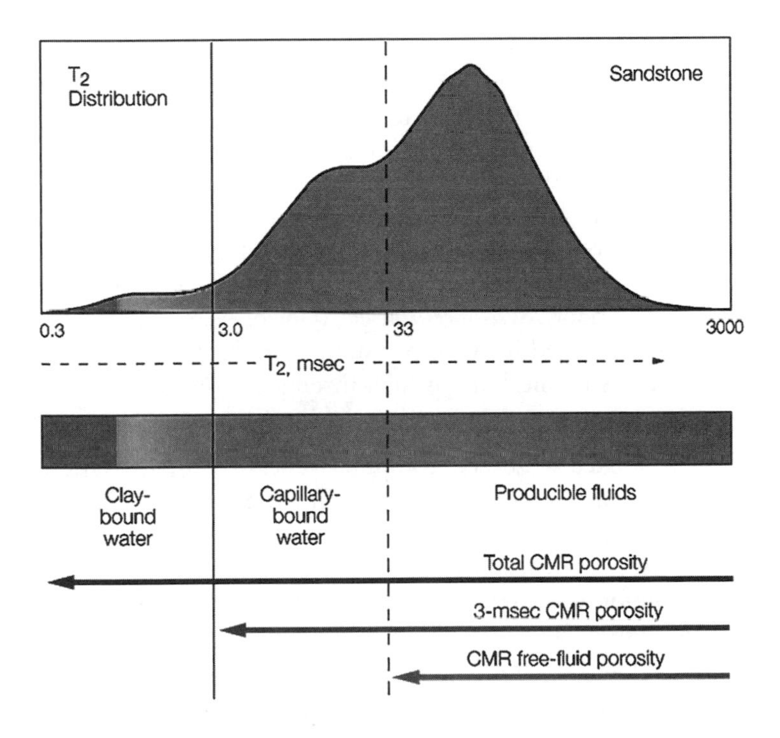

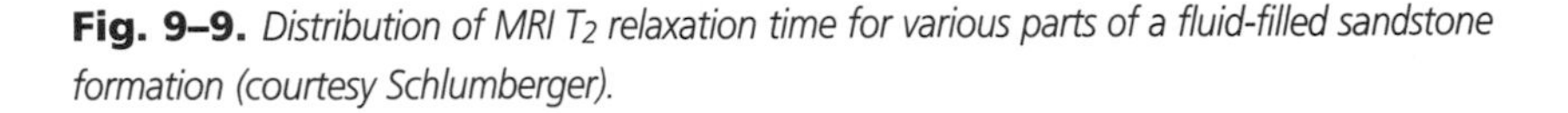

Fig. 9–9. *Distribution of MRI T_2 relaxation time for various parts of a fluid-filled sandstone formation (courtesy Schlumberger).*

Interpretation

The shape of the T_2 distribution curve is affected by the size of the pores. The smaller pores have lower permeability and occur earlier on the T_2 distribution curve. Clays have relatively high porosities, but permeabilities close to zero because of their very fine grain size and therefore very small pore size. The water contained in the clays is unable to move because of the lack of permeability. The clay bound water is essentially V_{sh}. The irreducible formation water (BVW_{irr}) is the formation water bound to the sand grains by capillary attraction. This water is not free to move because the pore size is still too small. The free-fluid portion of the porosity is characterized by larger pores. This fluid fraction is free to move and be produced.

If we add all of these fluid portions together, we get the total porosity. If we remove the clay-bound water portion, we have the effective porosity. These porosities can be determined independently of the lithology. The different regions in the T_2 distribution curve are determined by a T_2 cutoff that varies somewhat from one formation to another. In addition, permeability can be calculated using the NMR porosity and the T_2 curve.

Figure 9–10 shows GR and caliper curves in the left part of track 1. The T_2 distribution waveforms are shown in the remainder of track 1. The remainder of the log is divided into three tracks to the right of the depth track. Track 2 is a logarithmic scale of the dual laterolog resistivity and the permeability. Track 3 shows the neutron, density, and MRI porosity; track 4 is the bulk volume analysis. The top of the zone is at X468. The movable water begins at about X476 and increases rapidly for about 6 ft., after which it is nearly constant until the bottom of the zone. The log tells us that if we perforate below about X474, we will get a lot of water along with the oil. The well was perforated in the top 6 ft. and produced water-free oil. Note that in track 2 the permeability is high throughout the zone.

We have looked at several computer-generated interpretations. These were interpretations of formation resistivities, density and neutron porosities, natural gamma-ray radiation, and caliper curves. The interpretations gave us water saturations, effective porosity, and simple lithology. This is still the most common interpretation made today. Nearly every well drilled has this type of interpretation made on it—either by computer or by hand.

We have also looked at two different ways to arrive at porosity and/or water saturation by newer logs that do not depend on traditional measurements for answers. However, logs can give us even more information. In the next chapter, we'll look at some specialized logs and learn what kind of information we can glean from them.

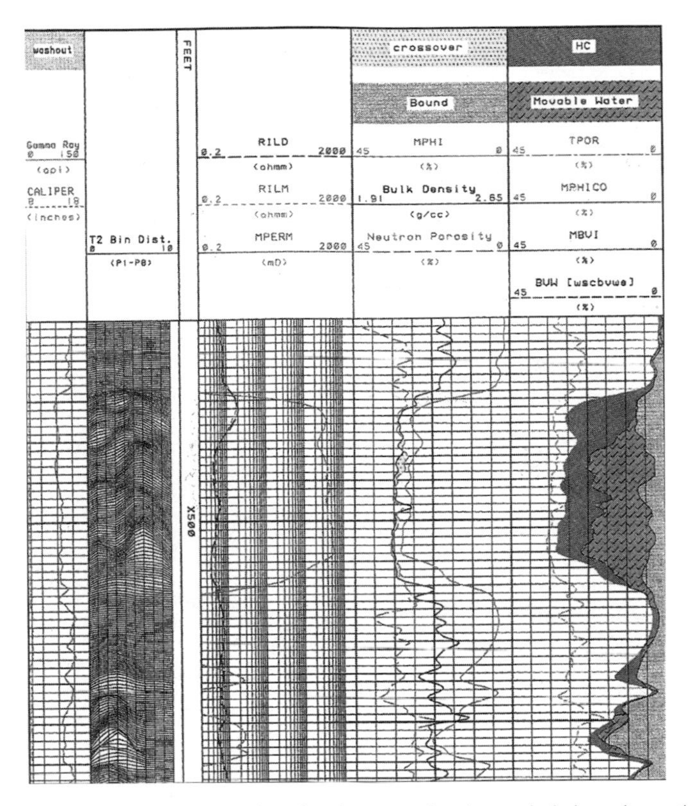

Fig. 9–10. *Interpreted MRI log combined with conventional open-hole logs shows the oil/water contact (courtesy Baker Atlas).*

10

Beyond Water Saturation

Electric logs, the original logs, were used to differentiate productive from unproductive intervals. At first the logs were just qualitative since they were totally new technology and little was known about their response or even exactly how they worked. Since that time, millions of dollars and man-hours have been spent in research by petroleum and logging company laboratories. The goal of the research has been to develop new and better ways to evaluate the earth's formations—especially those formations that might be petroleum reservoirs. At first, the research was directed at determining porosity, hydrocarbon content, and lithology. With the development of new sensors and powerful portable computers, the province of logging information has expanded astronomically. In this chapter, we'll look at some of the other ways well logs can be used to advance our understanding of subsurface strata.

WELLSITE SEISMIC TECHNIQUES

When a wildcat well is drilled, the location is often picked based on extensive seismic work. A seismic map reveals the possibility of a structure favorable to trapping hydrocarbons. Seismic work is expensive and time consuming. If the data can be made more useful or more accurate, the benefits could be great.

Readings are gathered by running lines of geophones along surveyed grids. A strong energy pulse—originally dynamite but now usually an air gun or a vibroseis unit—emits sound waves that travel along the surface of the earth and also into the earth. When a wave encounters a change in formation type, the wave is refracted. Usually some of the energy is reflected back to the surface and some of it is transmitted deeper where it may reflect off a still-deeper formation. The energy reflected to the surface geophones is recorded versus time on the central data processing unit.

Since a large number of geophones are in use at one time, many data points are gathered with each *shot* or energy pulse (Fig. 10–1). By interpreting the different reflection patterns and times, the seismologists (geologists who specialize in seismic interpretation) are able to map geologic structures buried deep within the earth.

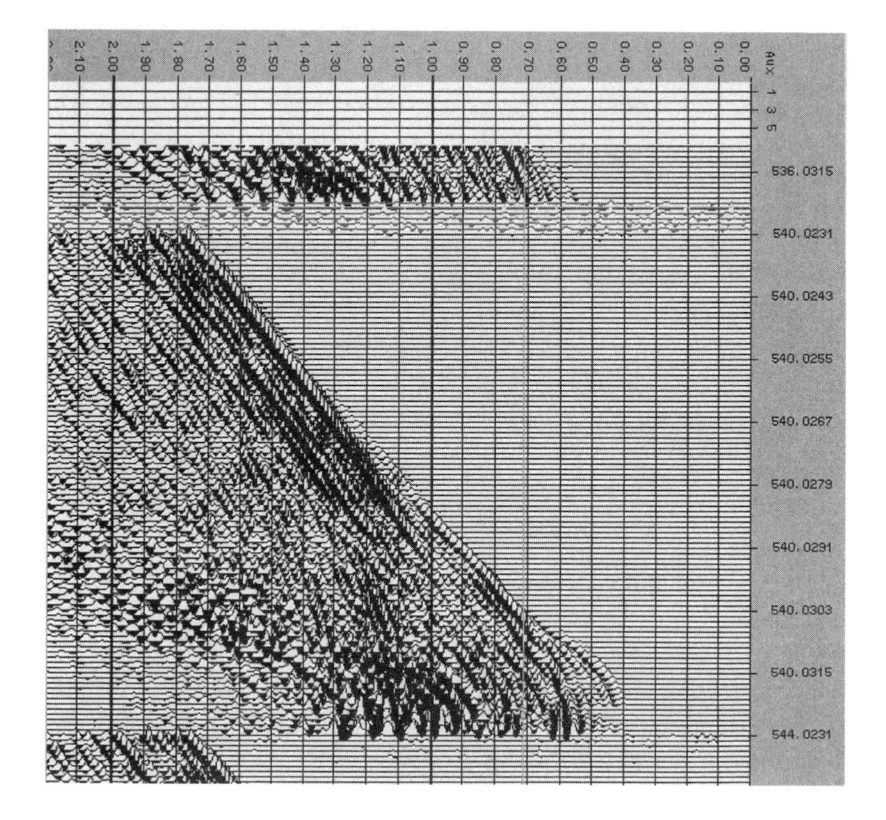

Fig. 10–1. *Sample seismic recording (courtesy Schlumberger). By interpreting the various sound reflections, geologists can find oil and gas traps.*

Check Shots

After the well has been drilled and logged, the surface seismic data can be made more useful by tying the well logs and the seismic data together. This is done by converting the well depth (in feet or meters) to a timeline (in milliseconds). The most accurate way to do this is with an integrated sonic log and check shots.

The sonic log traveltime is integrated by the computer from total depth (TD) to the top of the logged interval. The output is marked as time ticks in the depth column in milliseconds. If several different logging runs have been made at different depths, the integrated times are added together so that a total one-way traveltime from the surface to TD is obtained.

To make the check shots, a seismic geophone is run by the logging company and is stationed at various predetermined depths based on the formation changes. Figure 10–2 shows a typical setup for a check shot (also called a *well velocity survey*). An air gun, or vibroseis unit, supplies the energy pulse for the geophone. Several readings are taken at each interval. The time from the firing of the air gun to the first arrival at the geophone is measured; this is called the *one-way time*.

In Figure 10–2, geophones are positioned at points A, B, C, and D. The first arrival is represented by the first motion of the waveform at each level. By comparing the sonic time to the seismic one-way time, a correction curve called a *drift curve* can be made. The drift curve changes at each important formation interface. By relating the sonic integrated time to the seismic one-way time, a seismic time/depth line can be developed. This allows the seismic data to be scaled in depth for other nearby areas such as the next well location. The data can also be used to determine the thickness of the weathered zone and to make a weathered zone correction to the seismic data.

As can be seen in the figure, the sound waves are also reflected off deeper formations. These reflections, seen on the right of the figure and labeled "upgoing reflection," are used in the vertical seismic profile (VSP) log discussed in a later section.

Synthetic Seismograms

As the seismic wave travels through the different formations, the wave encounters formations of varying acoustic impedances. The acoustic impedance is a function of the formation density and the sonic velocity:

$$Z_a = \rho v$$

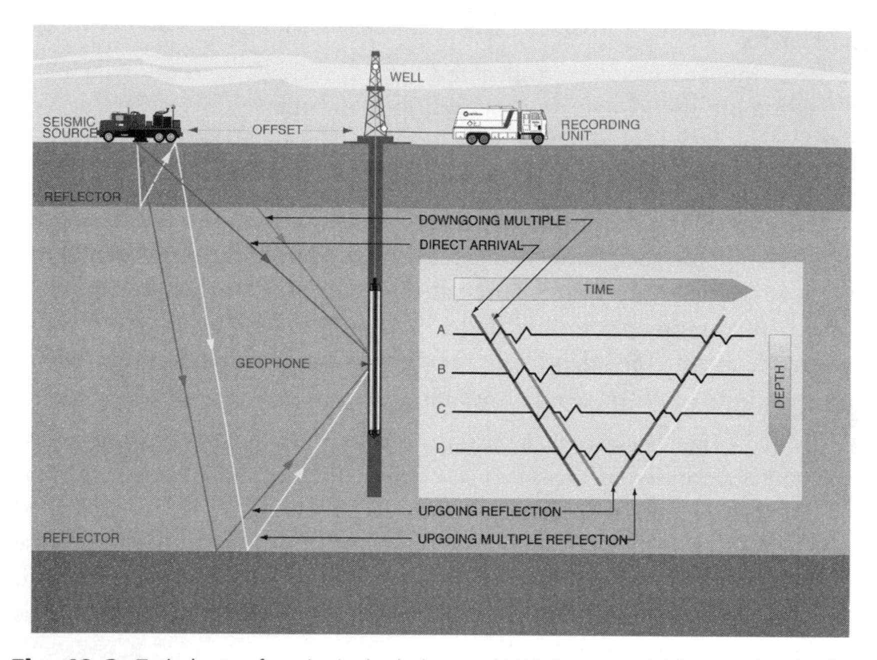

Fig. 10–2. *Typical setup for seismic check shots and VSPs (courtesy Halliburton). For check shots, only the direct arrivals are used. VSP surveys use both direct and reflected arrivals.*

where:

Z_a = acoustic impedance
ρ = formation density
v = acoustic velocity (the reciprocal of traveltime)

The formations' density and acoustic velocity are measured by the density and sonic (acoustic) logs. Therefore, we are able to calculate acoustic impedance at each change in the formation. The amount of energy reflected by the interface between two formations may be calculated from the acoustic reflection coefficient, R:

$$R_{1\text{-}2} = (Z_{a1} - Z_{a2})/(Z_{a1} + Z_{a2})$$

where:

$R_{1\text{-}2}$ = reflection coefficient at the interface between formations 1 and 2
Z_{a1} = acoustic impedance of formation
Z_{a2} = acoustic impedance of formation 2

The contrast in acoustic impedance determines the amount of energy reflected at each formation interface. This is easily calculated by the computer on a continuous basis. By introducing an artificial, idealized sound wavelet and combining it with the various coefficients of reflectivity, a *synthetic seismogram log* can be drawn. Several environmental corrections must be made to the sonic and density information before use. The insights gained by this presentation are of great use to seismic experts but hold little interest for us lesser mortals.

Vertical Seismic Profiling

The conventional surface seismic can only measure upgoing reflections, or *wavetrains*, while in VSP the geophone measures both upgoing and downgoing wavetrains. Figure 10–3 is a seismic recording taken in a wellbore. Shallow depth is at the right of the figure, with depths increasing to the left. Zero time (the time of the energy pulse) is at the top of the figure. Time increases toward the bottom. A large number of measurements are taken at 50 or more depths (*stations*). The computer separates the upgoing from the downgoing wavetrains. This information can then be used to study the change in the wavetrain with depth. In addition, the effect of the *weathered zone* (the top layer of the earth that is not yet compacted) is reduced since the wavetrain only passes through it once. The downgoing signals are the direct arrivals from the sound source through the formations and to the geophone. The upgoing signals have traveled from the sound source through the formations and then have reflected back up to the geophone.

VSP has several advantages:

- An actual seismic signal is recorded instead of a synthetic seismogram, making comparisons with the surface seismic data more meaningful.
- An exact correlation can be made between well logs and surface seismic data.
- Deep reflectors that may not be seen clearly on the surface seismic are recorded.
- Faults and other stratigraphic features near the wellbore are interpreted.
- The deconvolution operator for the surface seismic processing is determined.

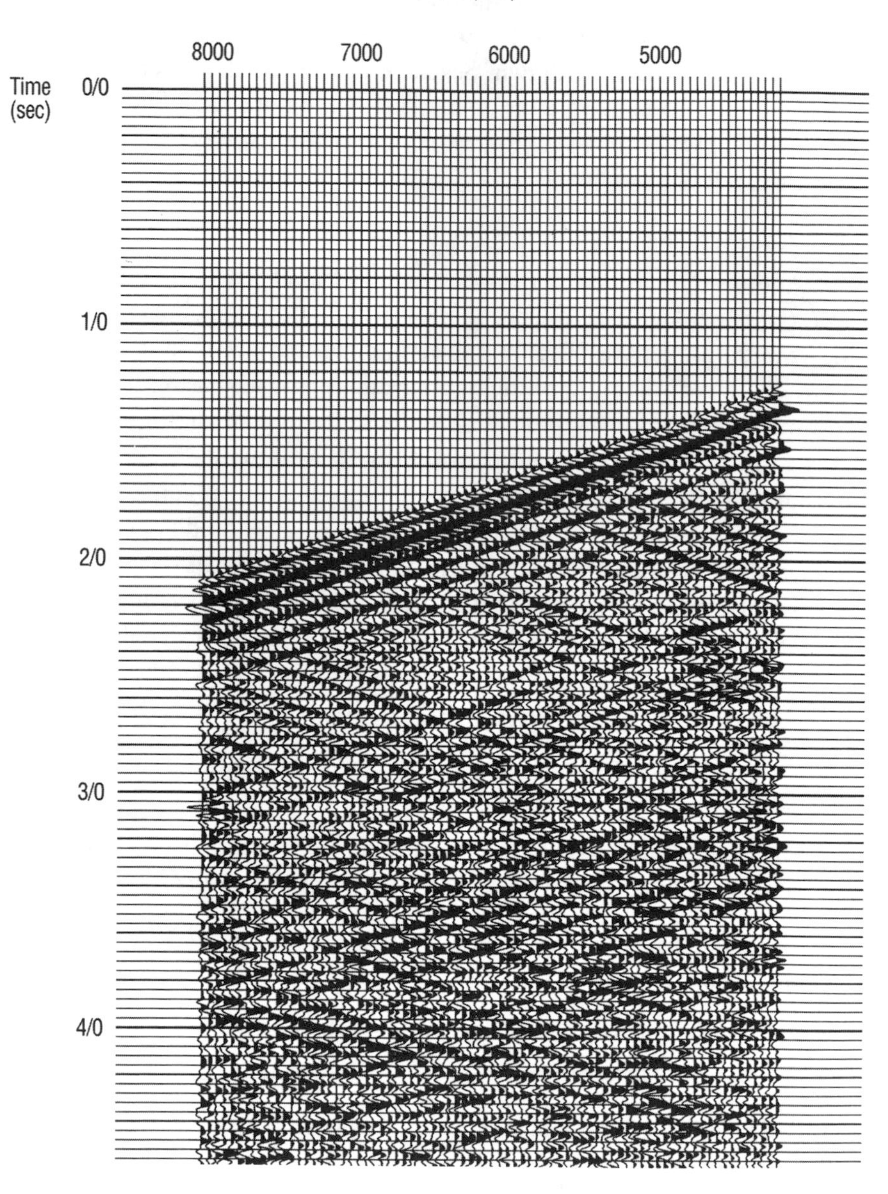

Fig. 10–3. *VSP survey (courtesy Halliburton). The information enhances the interpretation of surface seismic surveys.*

Normally the VSP is run with the sound source near the wellbore. This minimizes the distance the first sound arrival travels to arrive at the wellbore geophone. If the sound source is too far away, the arrival times must be corrected to true vertical depth (TVD). However, by moving the sound source farther away from the well, much more information can be gained about the nearby structure.

There are several ways to do this. First, we can use a stationary source at some fixed distance, called the *offset*, and take measurements with geophones at different depths in the well. Under favorable conditions, features can be detected laterally in the direction of the offset up to one-half the distance to the offset. This type of survey is called an *offset VSP*. Another technique is the *walkaway VSP*. In this system, the sound source is moved progressively farther away while recordings are made. The geophone is usually kept at the same depth, although an array of 50 or more geophones may be used if well conditions permit. This is usually done in cased wells to avoid sticking the geophone assembly. Offset and walkaway VSPs are often used to determine the presence of and distance to a salt dome.

DETERMINING DIP

One of the basic tenets of geology is that sedimentary rocks or beds are deposited horizontally, usually on the bottom of a lake, sea, or ocean. Therefore, the original dip of a reservoir rock is zero. However, this condition rarely persists. Because of geologic forces such as weathering and tectonics (forces within the earth that cause subsidence, mountain-building, earthquakes, etc.), the horizontal beds are tilted, faulted, compressed, and eroded. Often the formerly horizontal beds are shaped into *traps*, structures that inhibit the migration of hydrocarbons toward the surface. If we can identify the structure, we can develop the oilfield properly, i.e., we can choose accurate drill-site locations in optimum places.

By measuring the *dip* (the angle the formation makes below the horizontal) and *strike* (the direction toward which the bed is dipping) of the formations, we can infer a great deal about the processes that moved the beds to their present positions. Furthermore, the dip information helps identify the type of reservoir trap and the most likely location for the next well to be drilled. From dip information, we can determine many factors that help geologists and engineers determine how to develop a field.

To calculate dip, we need a large amount of information. Remember from high school geometry that any three points not on a straight line define a plane. If we can measure (1) the depth of a formation feature (a high resistivity value, for example) at three different points, (2) the displacement between the three points, and (3) the wellbore diameter, we can calculate the apparent dip using simple geometrical relationships.

To make these measurements, four microresistivity pads (instead of three) are pressed against the wall of the borehole (Fig. 10–4). The curves generated by these pads are correlated with each other to measure the displacement between each pair of curves (1–2, 1–3, 2–3, etc.). A caliper curve is recorded at the same time. (See Fig. 10–5 for a schematic of the dip tool in a deviated well.) To determine the strike or azimuth of the dip, the tool has a compass aligned with pad 1 to establish which direction the pads are facing when the measurements are made. From this information, we then calculate dip angle and direction.

Dips are calculated as though the wellbore were vertical; unfortunately, this is seldom the case. We must measure the angle the borehole makes to the vertical, called hole deviation or hole angle, as well as the direction the hole is going (hole azimuth) using pendulums and compasses within the dipmeter tool. With this information, we can correct the apparent dip direction and angle and find the true values of dip and strike.

Older dipmeter tools have three arms, the minimum needed. If one of the pads is floating or is reading incorrectly, there is not enough information to get one dip calculation. The four pads, using three pads at a time, give four calculated dips if all of the pads are making good contact with the formations.

Most of the tools in use today have four or even six pads mounted on individually articulated arms. Some of the four-armed tools have two electrodes on each pad. This means that up to eight correlation curves are recorded. Since only three curves are needed for a dip calculation, there is tremendous redundancy in measurement. The redundancy allows the dip information to be treated statistically. In addition, with two electrodes on each pad, very small features can be detected. This allows the calculated dips to show stratigraphic features such as crossbedding or direction of sand transport in addition to structural dip.

Computers can process a large number of dip calculations and allow a variety of interpretations. Relatively few points are used in the structural dipmeter presentation because structure changes slowly. For stratigraphic dips, a special dipmeter tool with six or eight correlation curves must be used; the larger number of correlation curves lets engineers calculate more dips. This is necessary to detect subtle changes in stratigraphy.

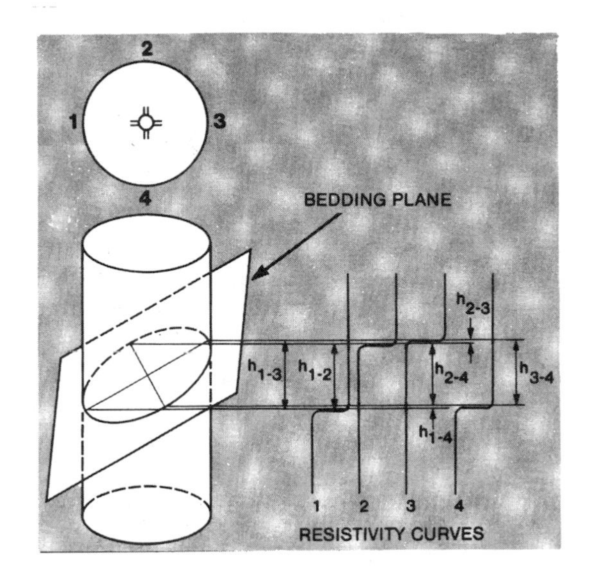

Fig. 10–4. *Principle of measuring dip (courtesy Schlumberger). Notice how the displacements relate to the intersection of a plane through a borehole.*

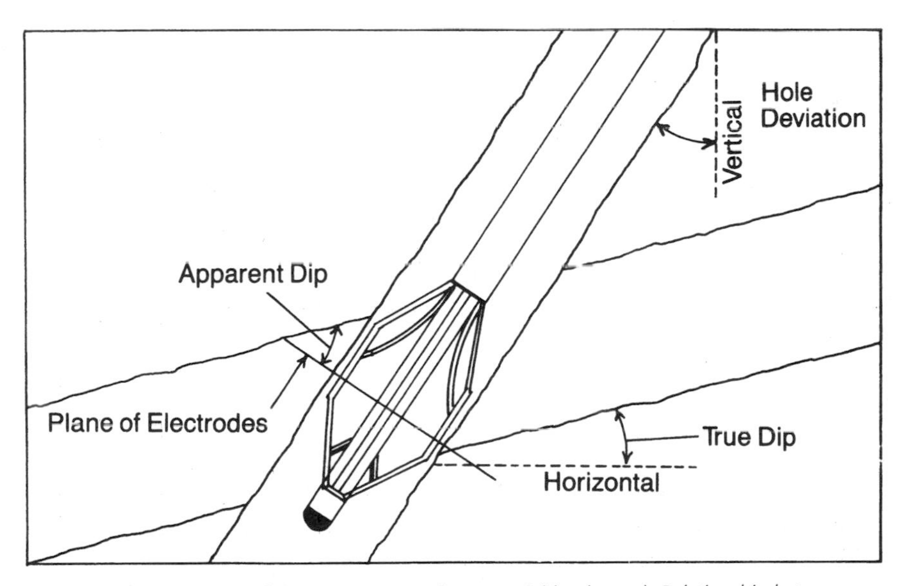

Fig. 10–5. *Geometry of dip measurement (courtesy Schlumberger). Relationship between apparent dip and true dip.*

Figure 10–6 is a presentation of a structural dipmeter log, often called a *tadpole plot* because the large "heads" and skinny "tails" of the plotted data look like tadpoles. Because of the detail in the presentation, we can discover not only large features such as anticlines, faults, or unconformities, but also subtleties such as sand transport or crossbedding. The geologist may refine the maps and more confidently pick the next well location with this information.

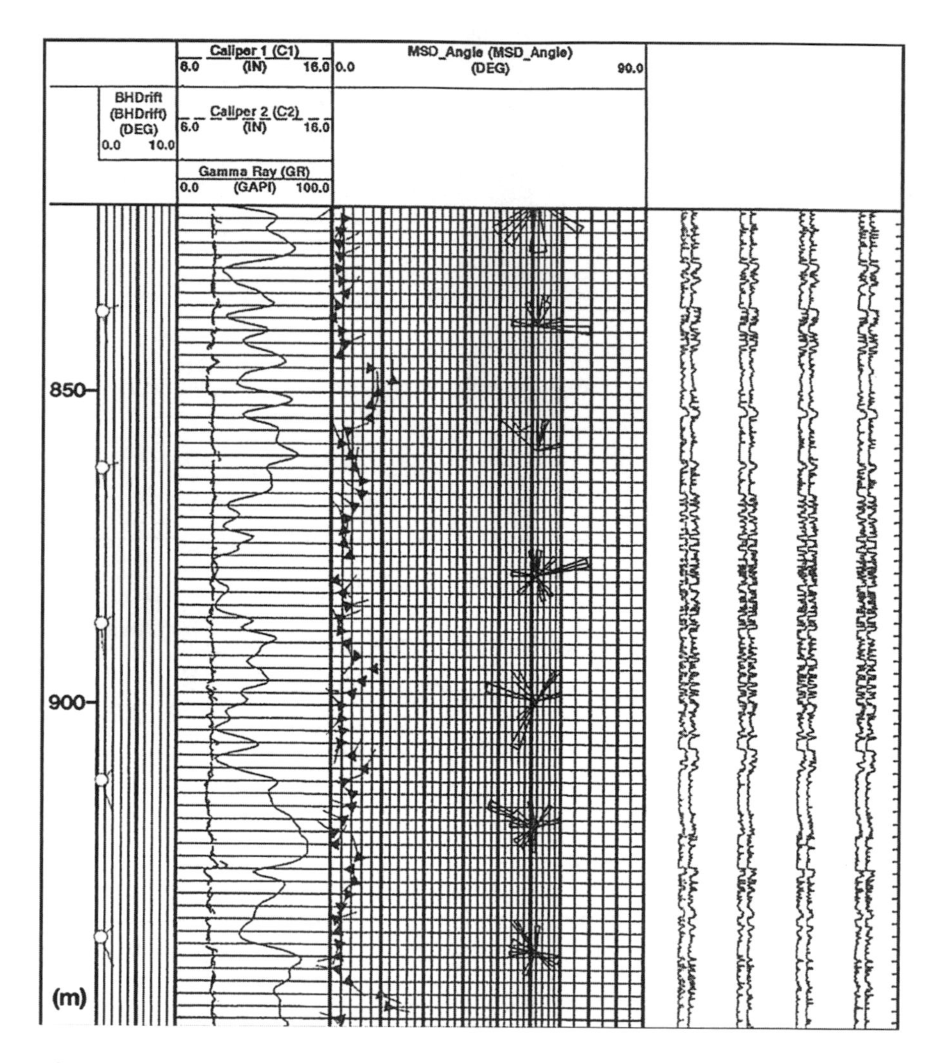

Fig. 10–6. *Formation micro imager (FMI) in dipmeter mode (courtesy Schlumberger). The tadpole plot is one way of presenting dip data. The angle of dip increases to the right on the graph; arrows (tadpoles) point in the direction of dip. The fan plots give an overall direction.*

Figure 10–7 shows a section of a processed dipmeter log. Notice the increased amount of data and alternative presentation of the information. In particular, note the pictorial, or *image*, of the wellbore in the column marked Stratim.

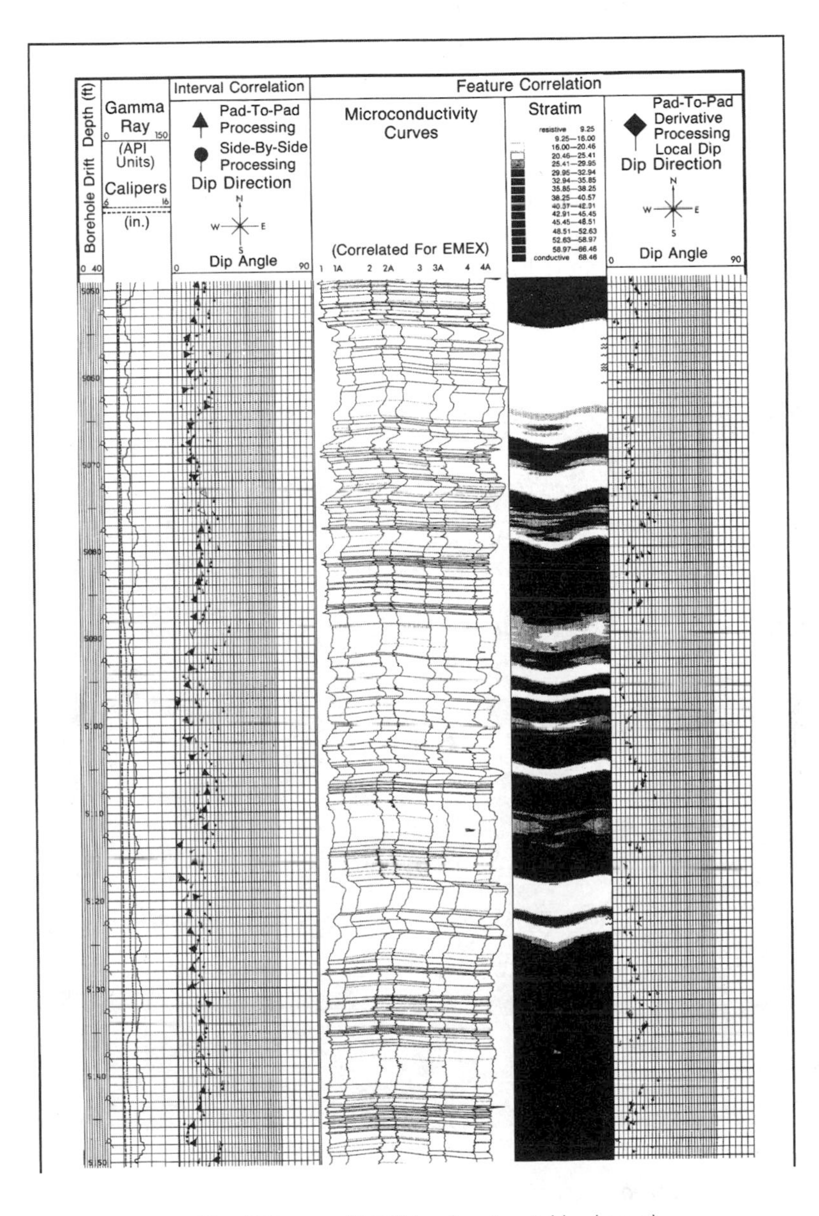

Fig. 10–7. *Processed Dual Dipmeter (SHDT) log (courtesy Schlumberger).*

Imaging Logs

One of the more recent developments, due strictly to greater computing power available, is imaging logs and the tools that make them. All major logging companies have imaging tools, which are a development of the earlier dip tools. One of the earliest imaging tools was essentially a dipmeter tool with a combination pad that had both dipmeter electrodes and an array of about 30 small electrodes. Although the coverage of the hole is only 20% or so depending on hole diameter, this Formation Micro Scanner (FMS) tool by Schlumberger is still in use.

Many of today's imaging tools still use a dipmeter tool frame, but a larger number of microresistivity sensors are mounted on from four to eight pads, depending on the particular design of the tool (see Fig. 10–8). The pad-mounted sensors are pressed against the face of the wellbore. A large percentage of the wall surface is swept by the pads, depending on the diam-

Fig. 10–8. *Simultaneous Acoustic and Resistivity (STAR) imager tool combines both resistivity and sonic measurements to give a detailed image of the formations (courtesy Baker Atlas).*

eter of the wellbore. The resistivity output of the sensors is color-coded such that higher resistivity is shaded with lighter or paler colors than lower resistivity. When all of the sensors are presented side by side, an amazingly detailed image of the wellbore is presented.

Figure 10–9 is a log section from Baker Atlas' STAR (simultaneous acoustic and resistivity) imager tool. The calculated dips are presented in the left track and measure between 70° and 80°. The actual measured resistivities for each pad are shown as a color image in the center track. Not all of the borehole wall surface is measured by the resistivity pads. The right track is the acoustic image. In addition to presenting a very clear picture augmenting the resistivity image, smaller features such as vugs, fractures, or crossbedding may be visible on the acoustic recording. The images are available in real time at the wellsite. These give a first look at fractures and secondary porosity features. Structural dips may also be calculated at the wellsite.

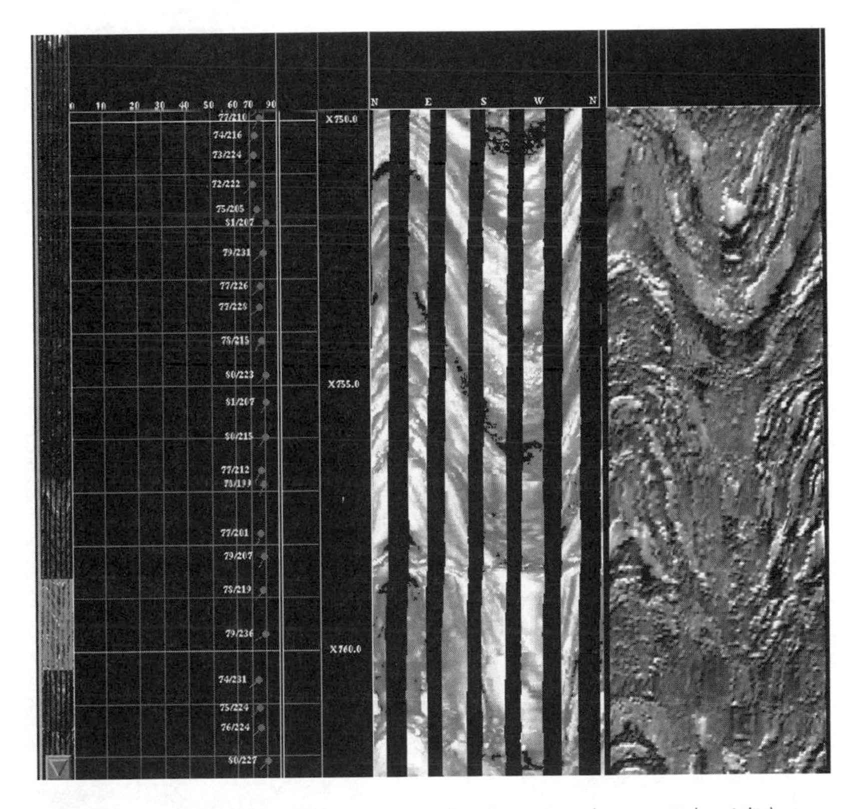

Fig. 10–9. *STAR imager log shows steeply dipping formations (courtesy Baker Atlas).*

A more detailed interpretation is made using a workstation at the computing center. Here, stratigraphic and structural dip features are calculated and compared to the image. The images themselves may be put on an expanded depth scale. An experienced analyst works alongside the oil company geologist to make the interpretation and describe the various features in detail. In Figure 10–10, note the expanded depth scale and the amount of information depicted on the log.

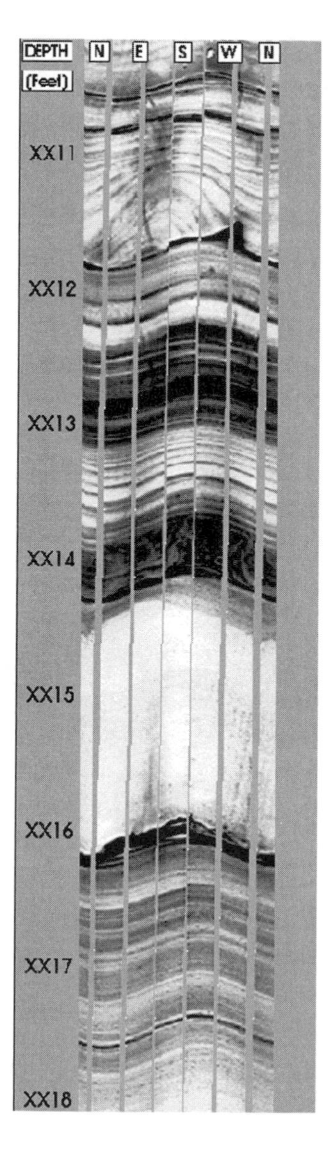

Fig. 10–10. *Workstation interpretation of an electrical microimaging (EMI) log (courtesy Halliburton).*

Acoustic imaging tools use a rotating transducer to give full 360° coverage of the wellbore. The advantage is more complete coverage of the surface of the borehole. Examples of this tool type are the Circumferential Acoustic Scanning Tool (CAST-V) by Halliburton and Schlumberger's Borehole Televiewer Tool (BTT). The Baker Atlas STAR is a combination acoustic and resistivity tool. As with the dipmeter-based tools, the main use for the acoustic imaging tool is for fracture identification, thin-bed recognition, secondary porosity determination, and possibly dip calculations with both structural and stratigraphic dips. In addition, acoustic tools may be used in casing to evaluate cement jobs and to inspect the casing itself. This is covered in Chapter 11.

Image presentation has become very popular, and its use has been extended to other logging tools. One such tool makes 12 deep resistivity measurements focused at different distances from the wellbore. The measurements help determine formation heterogeneities, thin-bed water saturation, and fractures complete with orientation. The images can be viewed at the wellsite, their color varying with the measured resistivity. While the image is similar to the image from microimaging tools, the detail is not nearly as great.

FORMATION TESTING

Formation testing is an important source of information on formations and their fluid content. While cores are commonly used to determine or verify porosity, permeability, lithology, and fluid type and saturation, formation tests are used to determine the ability of the formation to flow, formation pressure, and type of formation fluids. Formation tests may be made with tools run either on the drillpipe (*drillstem tester*) or on an electric wireline (*wireline formation tester*).

Drillstem Testing

The drillstem test (DST) is performed by running test tools that consist of open-hole packers and valves (Fig. 10–11). After the packer is set above the zone to be tested, a downhole valve is opened so the well can flow to the surface through the drillpipe. The drillpipe usually contains a cushion of fluid, but is not full.

The fluids in the formation are at reservoir pressure. When the test tools are opened to flow, fluid flows out of the rock, into the wellbore, and up the

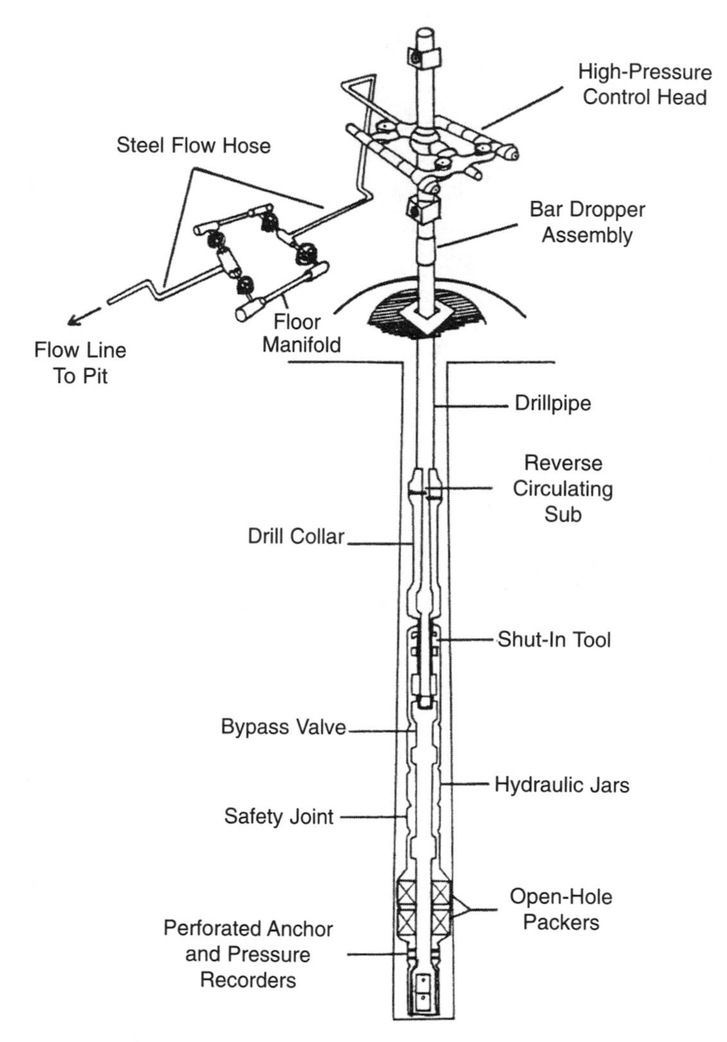

Fig. 10–11. *Schematic of tools used in a drillstem test. The open-hole packer seals the formation from the mud column and opens a flow path to the surface for formation fluids through the drillpipe.*

drillpipe. The downhole pressure gauge and recorder measure the pressure at all times. During the flow period, the pressure measured is called *flowing pressure*. After a certain flow period determined by the engineer, the test tools are closed or shut in. The maximum pressure measured during this time is called the *shut-in pressure*. As soon as the tools are closed, the well stops flowing and the pressure begins to build up from flowing pressure to reservoir pressure. Typically, the test tools are opened for a certain flow period, shut in for a period twice as long as the flow time, reopened, then shut in again.

If fluid reaches the surface, it is routed to a separator where the gas is metered and the liquids are measured and sent to a tank. If the well does not flow to the surface, the driller will be able to note any increase in fluid level when the pipe string is pulled. He can measure the level by counting the pipes that have been stacked in the derrick.

The chart or log of pressure versus time is called a *pressure buildup curve* (Fig. 10–12). Analysis of this curve helps engineers calculate reservoir permeability and pressure as well as formation damage.

DSTs have their drawbacks. They often take a long time to perform and so use up much valuable rig time. They are also expensive, especially com-

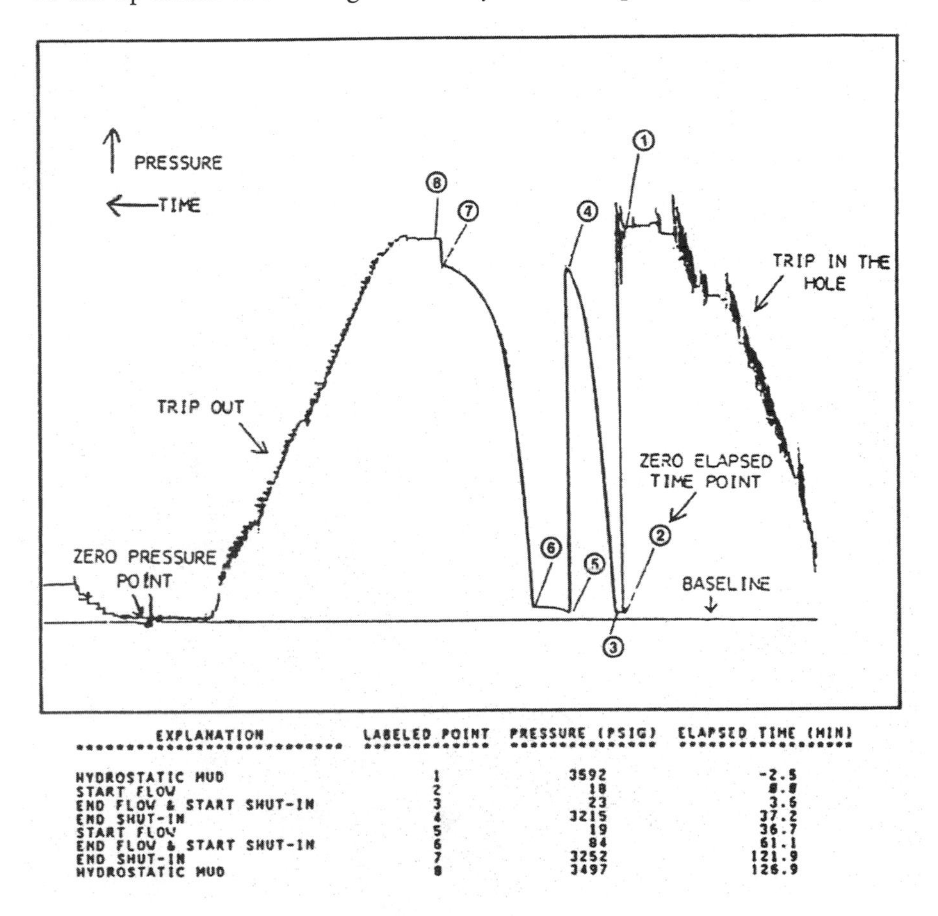

EXPLANATION	LABELED POINT	PRESSURE (PSIG)	ELAPSED TIME (MIN)
HYDROSTATIC MUD	1	3592	-2.5
START FLOW	2	18	0.0
END FLOW & START SHUT-IN	3	23	3.6
END SHUT-IN	4	3215	37.2
START FLOW	5	19	36.7
END FLOW & START SHUT-IN	6	84	61.1
END SHUT-IN	7	3252	121.9
HYDROSTATIC MUD	8	3497	126.9

Fig. 10–12. *Pressure buildup curve. See legend for interpretation.*

pared with wireline formation testing. And their results may be ambiguous in the following situations:

- If no fluid is recovered, the packer might not be set correctly or formation damage (clay swelling) might be the problem.
- If only mud filtrate is recovered, very deep invasion may be the answer. The zone could be productive or it could be wet.
- If drilling mud is recovered, the packer might not be making a good seal against the formation.
- If a long interval is tested, the location of the oil or gas might be difficult to pinpoint.

Wireline Formation Testing

Formation tests can also be made with a wireline tool. The theory of the wireline tester is simple: An empty sample chamber or jug is lowered until it is opposite the zone to be tested. A valve-controlled opening in the chamber, centered in a rubber pad pressed against the formation to seal out the mud in

Fig.10–13. *Wireline formation tester tool (courtesy Baker Atlas), showing the packer in the extended position. The sample chamber attaches to the bottom of the tool.*

the wellbore, opens and fluid flows in (see Fig. 10–13). A pressure gauge in the tool measures the flowing pressure. When the chamber is full, the pressure builds until it approaches reservoir pressure. Then the tool is closed, the seal to the formation is broken, and the tool is pulled back to the surface. The contents of the sample chamber are measured and analyzed, and the pressure buildup curve is used to calculate permeability and reservoir pressure.

Most tools today can be reset as often as needed to form a seal or to measure pressure only. The fluids usually flow through a metal probe or flow tube that is forced into the formation. In low-permeability formations or where fractures are suspected, a dual packer tool may be used that isolates a section of formation several feet long. This gives a much larger flow area and is more likely to recover a fluid sample. Samples are generally limited to one or two trials before the tool is pulled from the hole.

The wireline formation test tool is limited because the sample size is so small. In addition, the permeability measured by the probe is very close to the wellbore. This may not represent conditions deeper in the formation. Nevertheless, the wireline tester can be of great value when several zones must be tested, when a reservoir pressure measurement is needed (to check for a depleted reservoir, for example), or when permeabilities are high and a full sample chamber is likely to be obtained. With a full sample chamber, the type of fluid recovered and an estimate of permeability can be ascertained.

FORMATION CORES

While not a log in the strict sense, formation cores are an established way of evaluating the rock properties such as shaliness, type of shale, porosity, grain density, lithology, residual oil saturation, and permeability. It is similar to a doctor taking a biopsy to see exactly what is present. The cored formation is still the benchmark with which well-log responses are compared.

A core is taken when a formation of interest is about to be penetrated by the drill bit. Usually, the driller drills into the formation a foot or two because the exact depth of the formation is not yet known. The driller can often determine the top of the formation by the change in the bit penetration rate. This is called a *drilling break*. The cuttings that come up in the drilling mud are examined by the mud logger to verify that the objective formation has been found. The drillpipe and bit are pulled from the hole and the coring tools are run.

A hollow core barrel with a doughnut-shaped bit is attached to the end of the drillstring. Inside the core barrel is a rubberlike sleeve to help retain formation fluids and a mechanism to prevent the core from falling out of the core barrel. Some core barrels are designed to retain the core and fluids at reservoir pressure, but this is very expensive, requires special handling, and is not the usual case. After the 10- to 30-ft. core has been cut, the equipment is pulled out and disassembled. The core is usually removed from the core barrel at the well. The sections are inspected and labeled, a cursory examination and a few simple tests are made, and then the core is boxed and transported to a core laboratory for additional analysis. The analysis made on the cores is much the same as for conventional cores. Porosity, lithology analysis, grain density, clay content, and hydrocarbon shows can all be determined.

In addition to the conventional cores, small cores may be taken from the side of the wellbore with a tool run on the logging cable. This 50-year-old method uses a hollow bullet fired into the formation at operator-selected levels. The bullet is retrieved by small cables attached to the core gun. The cores taken are ¾–1 in. in diameter and up to 1½ in. long. Usually 30 or more cores are taken during one trip into the hole. On the surface, the cores are removed from the hollow bullets, placed in sample jars, and labeled. A cursory examination is usually made immediately by the wellsite geologist or engineer. The cores are then sent to a laboratory for a more detailed analysis.

A newer method of taking cores on a wireline is with a small rotary coring tool that cuts a core from the side of the wellbore (Fig. 10–14). Up to thirty cores measuring 1 in. in diameter and up to 1¾ in. long can be recovered on one trip. The main advantage of cut cores is no percussion damage, such as fracturing, unlike cores that are shot into the formation. After recovery, identification, and labeling, the cores are sent to a laboratory for analysis.

Cores taken on wireline have definite advantages over conventional coring:

- The interval to be cored can be determined after looking at the logs.
- The depth control is very precise.
- A large number of cores can be taken from a wide range of formations.
- If a core is missed, another try can be made. With conventional cores, they must be gathered on the first try.

Fig. 10–14. *Rotary sidewall coring tool (RSCT) (courtesy Halliburton). The hollow drill cuts a small rock sample from the formation. Operation is controlled from the surface.*

- The cost is much less than conventional coring.

LOGGING MEASUREMENTS WHILE DRILLING

At the end of the 20th century, the oil industry went through extreme consolidation. The price of oil and gas was very low. To operate at a profit, drilling and production practices had to be as efficient and cost effective as possible. Horizontal drilling techniques and other high-angle drilling programs are results of this drive for more efficient and productive wells. However, there are some obvious problems that a high-angle well presents to the logging industry.

The first is like the adage, "You can't push a rope." Logging tools cannot be run to the bottom (or end) of a horizontal well on a wireline cable. Once the angle of the wellbore is more than about 60°, friction overcomes the force of gravity and the tools stop. High-angle wells have been drilled offshore and in other locations for years. Techniques to log the wells have been developed, but they are not always satisfactory; the logging tools often end up stuck in the well. The logging program may then be cut short, resulting in less information than desirable.

Starting in the 1980s, a system of logging was developed that used logging tools incorporated in the drillstring. Development has been rapid over the last few years, with better data transmission systems and more measurements being made available. The basic *measurement while drilling* (MWD) system is a collection of tools built as a group of modules on the primary steering tool. The system provides the directional data necessary to steer the bit to its target and also obtains practically any logging data available from conventional logging tools: neutron and density porosity, resistivity, acoustic waveforms, weight on bit, and hole direction and angle.

The data are transmitted to the surface by pulses through the mud column. The pulses are decoded at the surface and displayed on a workstation. Then the information can be transmitted by phone line to the home office if desired.

One of the main advantages of MWD measurements is that formation information is obtained no matter what happens to the well. In exploration drilling, such as offshore, the main point of a well may sometimes be the log information. The MWD system ensures the information regardless of hole conditions or hurricanes, for example.

LOGS FOR AIR-DRILLED HOLES

In some parts of the country, wells are drilled using air as the circulating fluid rather than mud. This procedure is possible when formation pressures are low, the formation is tight (low porosity), and water is not present. Drilling with air is usually faster and cheaper than drilling with mud. However, the lack of a conductive mud system rules out the use of any logs that require a conductive fluid, such as microlog, SP, laterolog, or compensated neutron log.

The usual logging suite for these wells is an induction log (only one resistivity measurement is necessary because there is no invasion), a compensated density porosity log, an epithermal neutron porosity log, and often a temperature and noise log.

Epithermal Neutron Log

The epithermal neutron log is similar to the regular neutron log except that it detects high-energy epithermal neutrons. The source and detectors are mounted on a skid pressed against the side of the wellbore, just like a density tool. The log is used in air-drilled holes because it can measure apparent porosity in air, unlike the compensated neutron tool.

Temperature Log

The temperature log measures the changes in the temperature of the air or gas in the wellbore as the tool is lowered into (not raised from) the well. Normally, temperature increases with depth; however, when gas enters the hole from the formation, there is a cooling effect due to adiabatic expansion (explained more in Chapter 12). Cooling anomalies are of interest to engineers because they indicate gas entry. The temperature log must always be run first, after the wellbore temperature has had a chance to stabilize.

Noise Log

The noise log is essentially a microphone that amplifies any noise or sounds detected in the wellbore. The frequency of the sound encountered is recorded. Under optimum conditions, the type of fluid entering the wellbore from the formation, gas or liquid, can be determined from the frequency.

THROUGH-DRILLPIPE LOGGING

Sometimes it is impossible to lower conventional logging tools to the bottom of the well. The reasons might include swelling shales that bridge off (block or plug) the hole, ledges, poor mud properties, lost circulation zones, or high-pressure gas zones. For whatever reason, no conventional open-hole logs are obtained. In this case, there are only two alternatives: to run casing without seeing any logs (and then run a porosity device through the casing) or to run whatever logs are possible through the drillpipe. Usually, the geologist or drilling engineer decides to run the logs through the drillpipe.

Through-drillpipe logging is a technique rather than a tool. The drillpipe is first run in the hole open-ended (without a bit). The pipe is run past the obstruction and then is suspended by the slips at the rig floor. The logging cable is rigged up, and the tools are passed through the drillpipe and hopefully to the bottom of the well. The log is then pulled in the usual manner.

It is important to record the log while going in the hole as well as coming out. In a worst-case scenario, a down log may be all you get. Naturally, the logs available are those that will fit through the drillpipe (see Table 10–1).

Tool	Diameter, in.	Measurement
Induction electric	2¾	Resistivity
Density	2¾	Porosity
Compensated neutron	2¾	Porosity
Sonic	1 11/16	Temperature
Temperature	1 11/16	Correlation, lithology
Thermal decay tool	1 11/16	Porosity, sigma

Table 10–1. *Through-drillpipe logging tools*

These logging devices are limited in the variety of measurements available and often in the sophistication of the measurement. The resistivity measurement is usually a deep induction curve and a short normal curve. Although the density tool is compensated for mud cake, it does not have a lithology curve (photoelectric absorption index) or a caliper measurement. The thermal decay tool is normally a cased-hole device but may be used in an open hole. Cased-hole tools are discussed in the next chapter.

11

Completion Logs

Once a productive zone is found and the decision is made to complete the well, many problems must still be overcome before the oil, gas, or money starts flowing. For example, how can the productive intervals be isolated from the water zones or the low-pressure zones? In addition, the sides of the wellbore must be kept from collapsing or falling in and blocking the well so fluids will be able to flow. It didn't take the old-time drillers long to figure out they needed a strong liner for the wellbore. They called it *casing*.

Casing was first used in oil wells in the late 1800s. The first casing was wooden, fastened with steel hoops much like barrels. It wasn't long before steel pipe was used in place of wood. The advantages were obvious: the steel casing lasted much longer, it could be easily connected into long lengths, and it was readily available. However, steel casing was heavy and had to be supported in the well. If it wasn't supported, it could fall to the bottom of the well and block the producing formation. (In the days before casing was perforated, it was suspended above the producing zone—a *barefoot completion.*) In addition, the mere presence of the casing did nothing to isolate the various zones in the well. Something more was needed.

E.P. Halliburton invented the oil well cementing industry in Oklahoma in 1920 and solved zone isolation and support problems. Although his cementing techniques and cement itself have been improved over the years, the idea of isolating and supporting the casing by filling the gap between the casing and the wellbore remains the same today.

CEMENTING THE CASING

After the decision is made to complete the well, casing is run. This is the beginning of a complex operation. First, the drillpipe is run back in the hole and the hole is conditioned for running casing. This generally means the mud is circulated and brought up to specifications so the sides of the wellbore are as stable as possible with no signs of collapse or bridging. Next, the drillpipe is pulled out of the hole and usually laid down on the pipe racks. The casing is picked up one *stand*, or section, at a time; measured; and connected by means of a *collar* (a slightly larger pipe that has female threads on each end) to the *joint*, or length of casing hanging in the *slips*—a mechanism that locks around a joint of pipe and suspends it from the rig floor. This process continues until all the casing needed has been run in the hole. At predetermined intervals, *centralizers* and/or *scratchers* are attached to the outside of the casing to help isolate a particular interval.

On the bottom of the casing string is the casing *shoe*. The casing shoe is a bull-nosed (rounded) *sub* (specialized piece of short pipe) 2–3 ft. long that helps the casing slide off ledges or push through obstructions as it is lowered into the hole. Inside the casing shoe is a machined restriction called a *seat* that receives, or seats, the bottom plug. The bottom plug is used at the end of the cementing operation.

Once the casing is in position, the mud is again circulated to clean the well. A preceding fluid, often an acid wash, is pumped just in front of the cement to help clean the mud cake from the formation face. Next comes the cement. Oilfield cement is called *neat cement* because all it contains is Portland cement and water. (Concrete contains sand and gravel as well as Portland cement and water.) The mixture of cement and water is called *slurry*. Various chemicals may be added to change the properties of the slurry—the weight of the mix, the water loss, the setting time (either accelerated or retarded), etc. The cement mixes are often highly complex and may be designed in the cement company's laboratory to meet the needs of a specific well.

The cement and the preceding fluid are pumped down the casing, through the casing shoe, and back toward the surface. After the correct amount of cement is mixed and started on its way to the bottom of the casing, the bottom plug is inserted into the casing and pumped down the hole by the following, or *displacement*, fluid. The bottom plug separates the displacement fluid from the cement and stops the fluid from being pumped out of the casing and ruining the cement job. The casing usually is not cemented all the way to the surface but just across the productive zones in addition to any other zones that might give problems, such as water-bearing formations or low-pressure *thief zones.* (Formations with high permeability and pressure below that of nearby formations are often referred to as thief zones because they "steal" production. The oil or gas flows into the thief zone instead of to the surface.)

Two-Stage Cementing Jobs

Sometimes productive zones are separated widely in the well. If the distance is too great, it may be more economical to cement the casing in two stages. If the cementing operation is a two-stage job, a second-stage sub is inserted in the casing string just below the upper zone. Inside the second-stage sub are sliding valves, called ports, as well as machined seating surfaces of different diameters.

Three different-diameter plugs are used in this cement job. The first plug passes through the sub without opening the ports. This allows the first-stage cement to be pumped out of the casing shoe. The ports stay closed throughout the first-stage cement job. After the first stage is completed and the bottom plug has been pumped down, an opening bomb, or plug, is dropped from the surface. (Cementing plugs used to be called bombs because of their shape. The name was changed after a cement company expediter was arrested for trying to ship a "bomb" on an airplane—or so the story goes.)

This plug seats in the bottom of the two-stage sub. When pump pressure is applied, the sliding ports open, allowing circulation to the surface once more. The cement is then pumped as in the first-stage job but with a different set of volume calculations. The top plug is pumped down by the second-stage displacement fluid. This third plug seats in the top of the second-stage plug and seals the casing.

Pressure is often held on the casing to be sure all the plugs stay in place until the cement cures. After the cement has cured at least 24–48 hours, the casing is cleaned out to the top of the casing shoe with the drillpipe and bit.

Measuring Volume of Cement

The actual pumping of the cement is a measurement process. The volume of fluid in the casing is known by multiplying the length of casing by the cross-sectional area of the casing (volume = length X area). The amount of cement to be pumped is known from calculations made from the caliper logs, the casing size, and the height to which cement is desired. The cement is mixed continuously as it is pumped downhole.

The volume of the fluid that displaces the cement is the same as the volume of the casing. The rig pumps are calibrated in volume per stroke. To pump the different volumes, the drilling engineer just counts the pump strokes and turns the proper valves at the right times. It sounds simple, and in theory it is. However, we all know that if something can go wrong, it will. Pumps can break down, tired people can make the wrong calculations, a feed-water tank may run dry at a critical stage, or a formation may fracture unexpectedly. For whatever the reason, cement placement is sometimes less than perfect.

In the following sections, we will look at logs that correlate the open-hole logs to markers in the casing; evaluate the zone isolation and the support the cement gives; determine the downhole flow rates and the type of production; detect corrosion and pitting in the casing; and evaluate the formations for porosity, water saturation, and lithology.

CORRELATION LOGS

One of the most important measurements we make with wireline logs is the depth of the formations. Formation depths, as measured by open-hole logs, are benchmarks. The open-hole log formation depth is used for all depth comparisons and correlations. This point may seem obvious, but it is important. As Einstein discovered with time and Nietzsche with morality, everything is relative. The driller has one set of depths from the drilling log, the mud logger records the same formations at somewhat different depths, the open-hole logs find a third set of depths, and the cased-hole logger's depths will undoubtedly be different from all of the above. Which depth is correct? The pragmatic answer is that it doesn't really matter. We just have to pick one and stick with it.

Normally, everyone's depths agree within a few feet, but that is not exact enough. The wireline companies claim an accuracy of 3 ft. in 10,000 ft. This figure of 3 ft. is for absolute accuracy. The repeatability of depth measurement is much better—approximately 6 in. or less. That is, the difference between depths measured by the same cable, in the same well, will be within 6 in. The depth discrepancy between subsequent logs should also be 6 in. or less.

This repeatability can be expected even between different logging trucks and different logging companies. If Schlumberger logs the open hole and Halliburton logs the cased hole, all of the formation depths should be within 6 in. of each other. The same is true if Halliburton logs the open hole and Schlumberger or a third company logs the cased hole. By correlating one log to the next log, we can obtain this type of accuracy.

After the open-hole logs are run, the formation depths are referred to the depths shown on those logs. If a productive interval is at 10,350–10,352 ft., we must be certain we are perforating that exact 2-ft. interval and not another interval 5 or 10 ft. away. Depth control is critical; it must be exact. Perforating a well in the wrong place may cost thousands of dollars in repairs. It may cost tens or hundreds of thousands of dollars in lost production if the error is never discovered. Obviously, correlating the cased-hole depth measurements to the open-hole log measurements is of primary importance. So, how do we ensure that the cased-hole and open-hole depth measurements are the same?

The steel in the casing affects all of the common log measurements. Steel casing is electrically conductive. This characteristic effectively shorts out any resistivity signals, so laterologs, induction logs, and electric logs do not make useable measurements inside the casing. Nevertheless, most radiation measurements are only reduced in strength, not eliminated or blocked altogether. Since we commonly run GR and neutron logs in an open hole, it makes sense to turn to one or both of these measurements to correlate between the open-hole and the cased-hole log depths.

The GR tool is normally chosen for the correlation log. There is usually a lot of movement on the GR curve, from high to low values and back. This quality makes it easy to get an exact depth match when we compare the cased-hole GR log to the open-hole GR log. If the GR log lacks sufficient detail for a good correlation, we can often use the neutron log. In this case, we would correlate the open-hole and cased-hole neutron logs to get our depth match.

Gamma-Ray/CCL Tools

We have to be able to make an exact depth match to the open-hole logs every time we go in the hole with wireline. Many wireline trips are made throughout the completion process and later in the life of a well. One way to always make an exact depth correlation is to run a GR log each time we go in the hole. However, GR tools are relatively fragile and must be run slowly. If the GR log can be related, or tied in, to some casing feature such as the collars, then the process can be sped up to save valuable rig time.

The casing collars connect the joints of casing. The collars can be detected by a tool with an arrangement of permanent magnets and electrical coils.

This casing collar locator (CCL) tool is very rugged, has no moving parts, requires no external power, and is extremely reliable—ideal for rugged cased-hole operations. By running a collar locator with the GR tool, we can relate the open-hole logs to a casing feature that is easily and quickly detectable.

The GR/CCL tool measures the natural radiation inside the casing while recording depth of each casing collar as it is passed. The cased-hole GR log is then correlated to the open-hole logs. The casing collar depths can be marked on the open-hole logs, providing a fixed reference point that is easily found in the casing. For example, the top of our formation of interest might be at 8356 ft., which is 27 ft. below the casing collar at 8329 ft.

Casing used to be uneven in length, with each joint ranging 38–42 ft. long. When the GR/CCL tool was run, there was usually an unmistakable combination of lengths between the collars. For example, the distances between collars near the productive interval might be 40, 38.5, 40.5, 40, 40, 37.5, and 39 ft.. With a series of collar lengths like this, it was very easy to make the proper correlation. However, today's computer-controlled casing mills turn out casing joints that are consistently close to 40 ft. Nowadays, a series of collar lengths might be 40, 40, 40.5, 40, 40, and 40.5 ft. There is no way to make a guaranteed correlation in this instance because the series of lengths between collars is not unique.

Sometimes it is possible to locate a shorter joint some distance up the hole—for example, a second-stage sub—or the operator can prepare a special short joint of casing by cutting a regular 40-ft. joint in half and rethreading the cut ends. Once the correlation is made and the depths are absolutely certain, these short joints can be detected in the casing string and the tool slowly lowered to the zones of interest, counting each collar on the way down. With the short joints, there is no chance for error due to uniform length casing. Although the short joints take advance planning and a slight expense, they pay for themselves in quicker jobs and worry-free correlations.

If a short joint of casing has not been run and there is no sub in the casing string that can be used for correlation, the cased-hole logging engineer must not complete the operation. He must pull the logging tool or perforating gun out of the hole, go back in with a GR tool, and make the correlation. A GR tool must be used on every trip in the hole if there are no reliable markers in the casing. Mistakes in cased-hole operations can often be very costly, or even impossible, to repair.

CEMENT EVALUATION LOGS

Cement isolates zones. What do we mean by *zone isolation*? How do we know the cement job has provided zone isolation? What are some of the conditions or problems that may exist in a less-than-perfect cement job? Let's examine the logs that can provide the answers to these questions—sometimes.

Zone isolation means that one formation is not in communication with another formation due to fluids flowing between the casing and the formation face. The cement prevents fluids from flowing either up- or downhole between the casing and the wellbore. The cement makes a hydraulic seal between the casing and the formation. In addition, it prevents injected fluids, such as acid or fracture treatments, from flowing up or down the hole behind the casing. In a sense, zone isolation depends on the conditions in a specific well. If the well will never be subjected to high-pressure treatments but only to formation drawdown pressures, then the cement job doesn't have to be as "good" as one that will be subjected to these stresses.

The cement bond log (CBL) was one of the first attempts at providing an exact answer to the zone isolation question. CBL tools are often an adaptation of open-hole acoustic or sonic logging tools. In its simplest form, the CBL uses a transmitter and two receivers (Fig. 11–1). The transmitter emits a pulse of sound energy that travels in all directions—through the casing, the cement, and the formation. Some of the acoustic energy is returned to the two receivers.

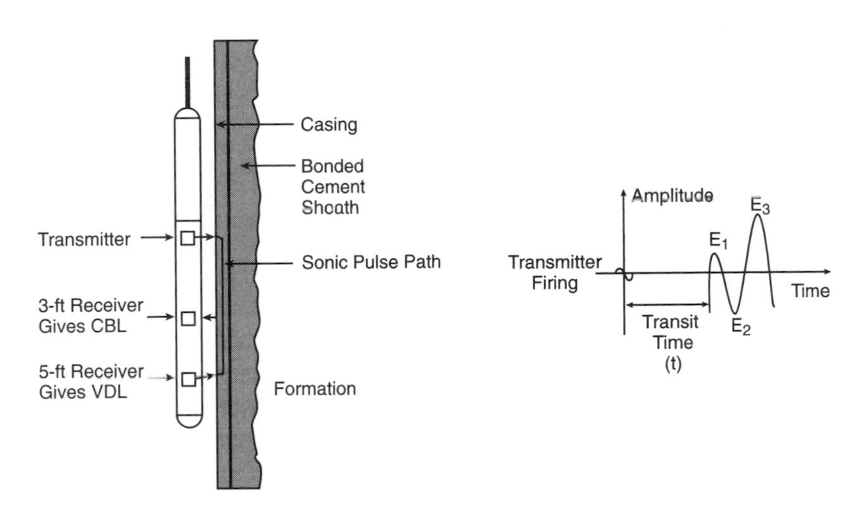

Fig. 11–1. *Simple cement bond logging tool (courtesy Schlumberger). By measuring the attenuation of the E_1 amplitude, the cement bond can be estimated.*

The first sound energy arrival (*first arrival*) at the top receiver comes through the casing. The sound-wave energy is represented by an oscillating signal in Figure 11–1. The circuitry in the tool measures the time and the amplitude, in millivolts, of the first arrival. The first positive signal is labeled E_1, the first negative arrival is E_2, the second positive arrival is called E_3, and so forth. The amplitude of the E_1 arrival is a function of the casing diameter, the casing thickness, the cement compressive strength, the percent of the casing circumference that is bonded with cement, and E_0 amplitude. It is a measure of the attenuation of the sound energy as it travels through the casing, indicating the effectiveness of the bond between the cement and the casing.

The concept of *attenuation* of sound in the casing due to cement outside the casing may be a little hard to grasp. Think of tapping a bell suspended in the air. The metal vibrates freely and the bell rings. If you put your hand on the bell and tap it, again it will still ring but not as loudly as before. The sound is attenuated. If the bell were coated with a thick layer of cement and tapped once more, the only sound would be the tap. The vibration would be almost completely attenuated.

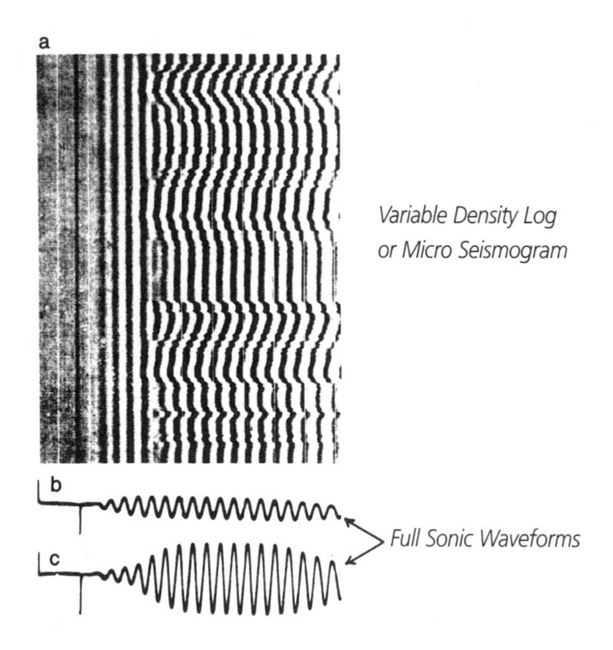

Fig. 11–2. *Full-waveform recording of the acoustic energy arrivals at the 5-ft receiver (courtesy Welex). The MSG, or VDL, is made by coding the positive waveforms (black) and the negative signals (white).*

In addition to the E_1 amplitude curve, the CBL tool usually has a full-waveform recording in track 3 (see Fig. 11–2). This image is variously called a Micro-Seismogram (MSG, by Halliburton) or a Variable Density Log (VDL, by Schlumberger). Both the VDL and the MSG are usually recorded with the 5-ft. receiver. (The term VDL includes the MSG for simplicity's sake. If there is a basic difference in the two measurements, the difference is noted.) The VDL image gives a qualitative indication of the cement bond to the formation. Strong formation arrivals, indicated by changes in the VDL image that correlate to formation variations, mean good cement-to-formation bond. Free pipe with no cement is indicated if the VDL shows strong parallel black-and-white stripes and high E_1 amplitude in track 2. To have good zone isolation, the cement must be bonded to both casing and formation.

Look at Figure 11–3, a GR/CCL/CBL/MSG. (Remember: MSG is the same as VDL, just different tool manufacturers.) The GR curve is used to correlate to the open-hole logs, the CCL is used to relate the GR to the casing collars, and the CBL/MSG evaluates the cement job. Look at the MSG (track 3) in the bottom log section. The waveforms correlate with the GR curve. Also notice how the amplitude curve (track 2) is generally low throughout this interval. The maximum reading is 5–7 mV in the middle of the zone and about 1 mV in the remainder of the interval. (The slightly increased readings opposite the productive interval are common. They may be attributable to gas infiltrating the cement.) The low amplitude and strong formation arrivals are indications of "good" cement.

The cement map located between T2 and T3 shows almost solid cement in the interval. Look at the casing collars in T1. Collars are shown at X760, X714, and X688. The distance between collars 1 and 2 is 46 ft., and the distance between collars 2 and 3 is 26 ft. The short joint is very useful for depth control on future logs.

Now look at the top section of the log. The amplitude curve is reading about 70 mv, and MSG has some early arrivals that are strong and do not vary with the GR or with depth. This indicates poor or no cement.

Example Problem. Figure 11–4 is a nomograph that allows us to make a quantitative interpretation of the cement bond by estimating the percent of the circumference of the casing that is bonded. The percent of circumference bonded is often referred to as bond index (BI). To use Figure 11–4, we need the following information:

- Casing thickness
- Cement compressive strength
- Casing size
- CBL amplitude in free pipe (uncemented zone)
- CBL amplitude in the zone of interest

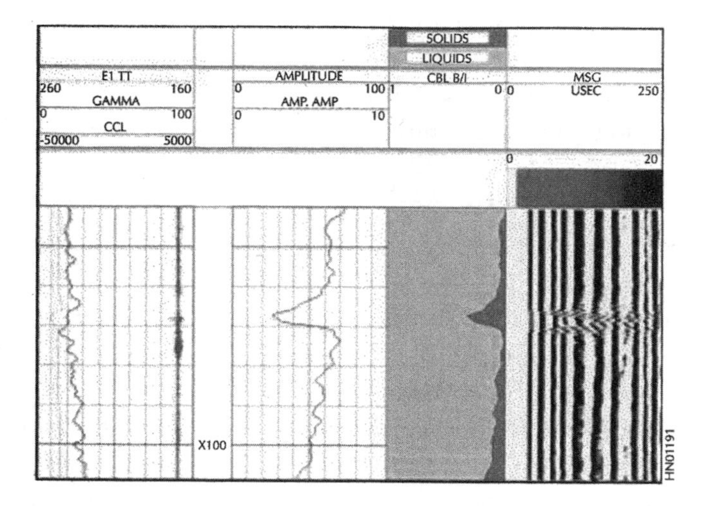

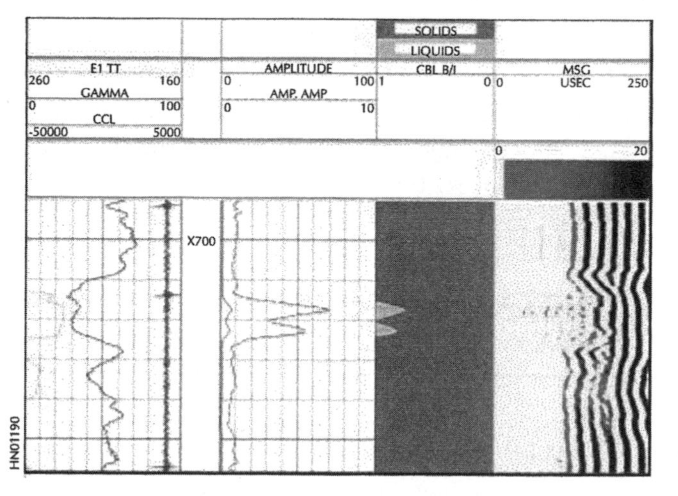

Fig. 11–3. *CBL good bond/bad bond example (courtesy Halliburton). In the bottom example, notice the MSG correlation to the GR log in track 2. In the top example, notice the strong parallel bands on the MSG typical of casing signals.*

First, find the casing outside diameter (OD) at the top left of the nomograph and draw a line vertically down to the Free Pipe curve. From this intersection, draw a horizontal line to the right to the CBL Amplitude line. If you are following the example on the figure, you are now at the FP point. Mark this point FP. (If you use 6-in. OD casing instead of 7-in., your FP point would be at 70 mV instead of 60 mV as in the example. If you are interpreting a well, use the casing size for your well.)

Next, find the casing thickness at the bottom left side of the nomograph. You can get this information, along with cement compressive strength, from your cement supplier's cementer handbook. In our example, we have 5⁄16-in. (8-mm) pipe thickness and 2000-psi cement. Draw a line horizontally from the casing thickness to the intersection of the appropriate cement strength line. Then, draw a vertical line from this intersection to the casing size line.

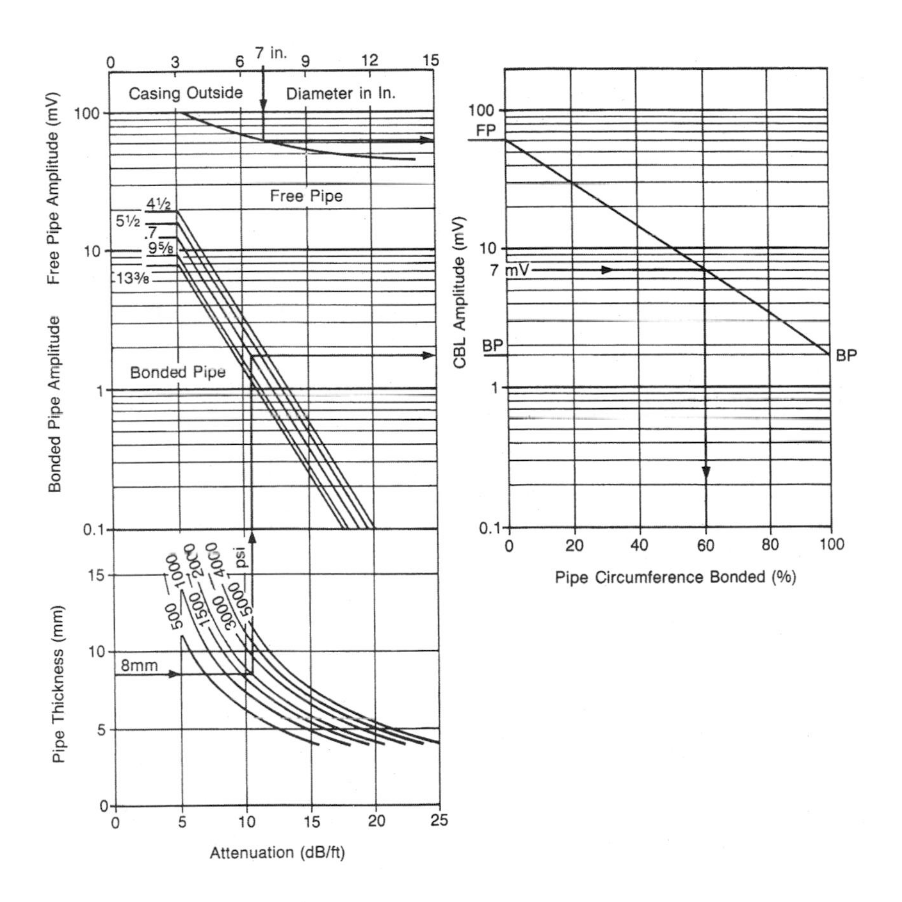

Fig. 11–4. *Nomograph for calculating bond index (courtesy Schlumberger).*

From here, go horizontally to the right to the CBL Amplitude line. Extend your line to the 100% line and mark this point BP, for bonded pipe. If you did everything right in the example, the point is already labeled for you. Connect BP to FP with a straight line. You have now calibrated the chart to interpret your well.

To evaluate a zone using the nomograph, read the amplitude curve on the CBL in the zone. Let's say you read 7 mV. Enter the nomograph on the CBL Amplitude line at 7 mV. Draw a horizontal line to intersect with your calibration FP/BP line. From this point, drop a vertical line to the Pipe Circumference Bonded Line. In the example, we get 60%, or BI = 0.6. When doing an interpretation by hand, we usually average the amplitude reading over several feet or until there is a significant change in the reading. We then make one calculation for this interval and draw the result on a copy of the bond log. Now, of course, the computer makes all the calculations and draws a bond index curve for us on a foot-by-foot basis.

Experience has shown that the number of feet of well-bonded casing is a function of the casing size. In other words, the larger the casing diameter, the greater the number of feet of well-bonded casing needed. Figure 11–5 relates casing size to the minimum number of feet of cement with a BI of 0.8 or greater that is needed to achieve zone isolation in most cases. For example, 5½-in. casing needs 5 ft. of interval with a BI ≥ 0.8; and 7-in. casing requires nearly 10 ft. of a BI ≥ 0.8 to be assured of zone isolation, according to Figure 11–5.

Conditions That Affect CBL Interpretations

The presence or absence of the designed compressive-strength cement is the primary factor that affects the E1 amplitude on the cement bond log. However, other conditions can affect the interpretation of the CBL: channeled cement, gas-cut cement, a microannulus between the casing and cement sheath, thin cement sheaths, or fast formation arrivals.

Channeled cement. Several conditions may lead to a channeled cement job. One of the most common causes is *eccentered casing*. This means the space between the casing and the wellbore is not uniform. Often the casing contacts the formation on one side. In this case, the cement is unable to completely surround the casing. A channel is left on each side of the line of contact of the casing with the formation. (This type of channel is almost impossible to fix.) The cure for this problem is prevention: centralizing the casing with 3- to 4-ft.-long mechanical basket-type devices (centralizers) that fit over the outside of the casing and are used above and below zones of interest.

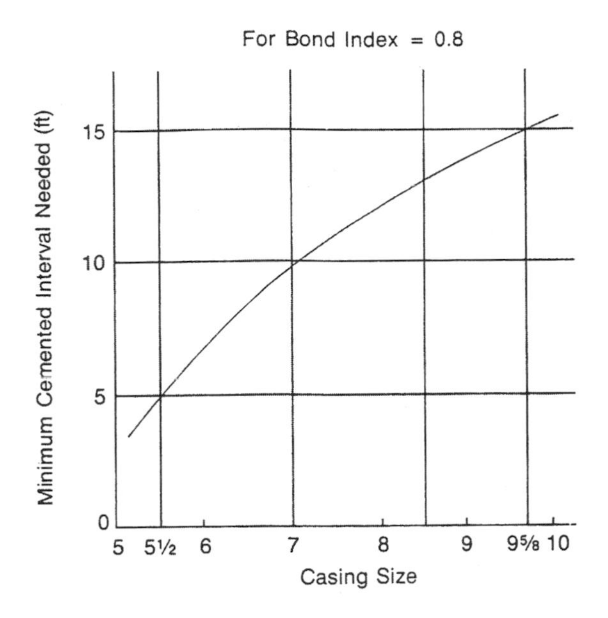

Fig. 11–5. *Nomograph used to calculate the number of feet needed for zone isolation (courtesy Schlumberger).*

Another cause of channeling is formation fluid flow before the cement has cured or set up. This can happen if the preceding fluid that was used to flush the mud and clean the formation face was not heavy enough to hold back the formation fluids. In this type of channeling, the well flows into the annulus. The flow may reach the surface, or it may be diverted by a low-pressure formation that "drinks" the fluid.

Channels appear as a higher amplitude reading on the CBL. The VDL image may show casing arrivals as well as formation arrivals.

Gas-cut cement. Normally, cement has a permeability of about 0.001 millidarcies (md). This is essentially zero, or impermeable to the flow of liquid or gas. However, as the cement loses water to the formation and begins to cure, the pore pressure within the cement drops. In a high-pressure gas zone, the gas may invade the cement as the cement pore pressure decreases. The gas-invaded cement is called *gas-cut cement.* Gas-cut cement can have a permeability of nearly 5 md. The cement allows gas migration although it still supports the casing mechanically. Gas-cut cement is difficult to identify on the E_1 amplitude curve. The VDL image may appear washed-out or gray in appearance.

Microannulus. A microannulus is a very small gap between the casing and the cured cement. It normally is too narrow (0.002–0.004 in.) to allow fluid to flow. Sometimes it is necessary to hold pressure on the casing due to mechanical problems or well conditions during or immediately after the cement job. If the pressure is held until the cement cures, a microannulus is often the result. Another cause of a microannulus can be due to displacing the cement with heavy drilling mud. When the drilling mud is replaced with a lighter completion fluid, such as saltwater, the difference in hydrostatic pressure can cause a microannulus. A microannulus may be created any time pressure is applied to the casing and then released. Cased-hole drillstem tests and remedial cement jobs often cause a microannulus condition.

Zone isolation and pipe support still exist with a microannulus. The problem is recognizing and interpreting the microannulus. The amplitude curve on the CBL will show high values—often close to free pipe readings. The VDL generally has strong casing arrivals as well as formation arrivals. In cases where a microannulus is suspected, the proper procedure is to apply pressure to the casing and rerun the log. It is not necessary to reapply the same pressure that caused the microannulus; it is only necessary to verify that a microannulus exists. If the E_1 amplitude is reduced, and the casing arrivals are reduced in strength on the VDL, the microannulus is confirmed and the cement is assumed good (adequate).

Thin cement sheaths. If the cement sheath is thinner than ¾ in., the cement is unable to properly dampen the vibrations of the casing and the amplitude curve will read too high. This condition should be suspected any time the open-hole caliper log shows very thick mud cake on a formation or where the open-hole diameter is < 1½ in. greater than the casing size. Normally, the open-hole size is 2 in. or more larger than the casing size.

Fast formation arrivals. Low-porosity, dense formations such as limestones and dolomites have acoustic velocities faster than steel. The traveltime of steel is 56 µs/ft. Zero-porosity limestone has a traveltime of 49 µs/ft., and dolomite is 44 µs/ft. The acoustic energy from the fast formations arrives at the receivers on the cement bond tool before the casing signals arrive. This can only happen when the casing is well bonded; otherwise, the fast formation signals would be absorbed by the liquid in the space between the casing and the formation.

Fast formation signals are usually easy to recognize. The traveltime curve (track 1) records a lower value than in free pipe. The amplitude curve is not valid under this condition and may read either high or low values. When interpreting, disregard the amplitude curve over the interval of fast formations. The

VDL image will verify fast formation arrivals and a good cement bond by showing strong arrivals that are earlier than the casing arrivals. (The casing arrivals are shown in the free-pipe section of the log.) In addition, fast formation arrivals should be expected after looking at the open-hole porosity logs.

Pulse Echo Tools

CBL/VDL interpretation is complicated and somewhat subjective. In the mid-1980s, another type of cement evaluation measurement was introduced—the Pulse Echo tool (PET) by Halliburton and its competitive counterparts, the Segmented Bond tool (Baker Atlas) and the Cement Evaluation tool (Schlumberger).

Sound energy travels perpendicular to the axis of the casing and measures the attenuation due to shear coupling of the cement and casing. The technique emits a pulse of very-high-frequency sound energy (about 250–650 kHz) in the resonant frequency of the casing thickness. The casing is made to vibrate at right angles to its long axis. The vibrations of the signal are a function of the acoustic impedance of the fluid (mud or saltwater) between the tool and the casing, the casing, and the material (fluid or cement) between the casing and the formation.

The casing vibration dies out at essentially an exponential rate of decay. The changes in this exponential rate of decay are from changes in the acoustic impedance of whatever lies between the casing and the formation. (The fluid inside the casing and the casing itself are generally the same between measurements.) In the case of uncemented casing with saltwater both inside and outside, the casing vibrates for a relatively long time and the decay rate is slow. If cement is on the outside of the casing, then the vibrations are quickly dampened and the decay rate is fast. The difference in response is a function of the acoustic impedance of the cement.

The PET has eight transducers located in a double spiral around the tool such that a reading is made every 45°. The tool is centralized in the casing so the distance from each transducer to the casing is the same. In addition, a ninth transducer is used to measure the acoustic impedance of the fluid inside the casing. (Often there are changes in the density and composition of the fluid in the casing. These changes could cause a misinterpretation if not taken into account.)

Figure 11–6 compares a GR/CCL/CBL/VDL log (left) and an SBT log (right). The right side of the SBT log shows the output of each of the six transducers scaled in attentuation. To the left of the depth track is the

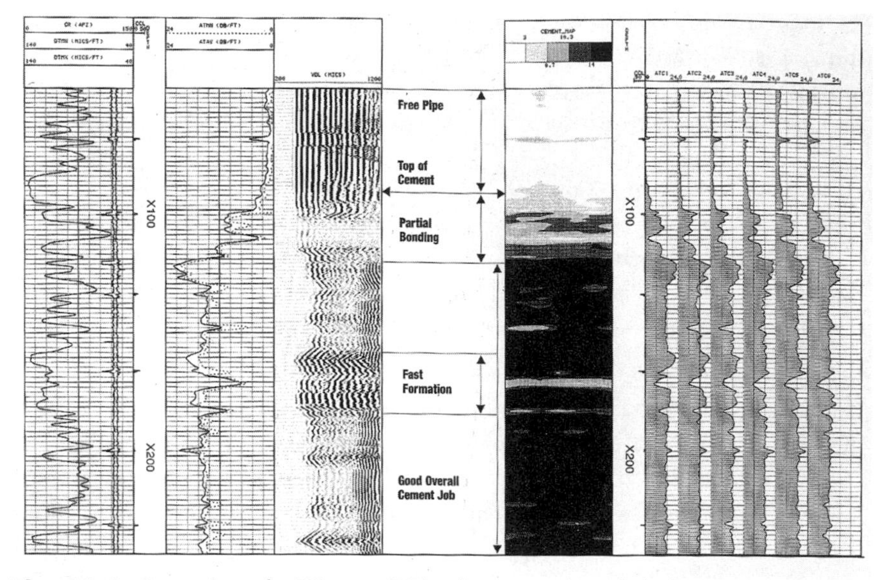

Fig. 11–6. *Comparison of a CBL to an SBT log (courtesy Baker Atlas). The cement map on the SBT is an interpretation of the six attenuation curves shown on the right of the log.*

cement map. In this example, the cement is uniformly good until the top is reached at about X100 ft. Compare the CBL and the SBT, and note the responses of the two tools in the different zones—no cement, partial bonding, fast formations, and good bonding.

An advantage of the SBT tool is that it is unaffected by microannuli. The main disadvantage of the tool is that it does not provide 100% coverage of the casing surface. The lack of coverage problem has been solved in the next generation of ultrasonic cement evaluation tools.

Ultrasonic Cement Evaluation Tools

The newer cement evaluation tools use the same measurement principle as the PET, but they have solved the spotty coverage problem by using one rotating transducer instead of six or eight fixed transducers. The rotating transducer gives 100% coverage of the casing surface. The coverage is so complete and detailed that it can be used for casing inspection as well as cement evaluation. A different transducer is required for each size of casing. The tools are run in different modes, depending on what is required of them—cement evaluation or casing inspection. Halliburton's tool is the Circumferential Acoustic Scanning Tool, CAST-V™. The equivalent Schlumberger service is the UltraSonic Imager, USI™.

Figure 11–7 shows a section of a USI log with an obvious channel revealed in the cement map (right side). The casing thickness and ID/OD are also shown.

The real advantage of ultrasonic cement evaluation tools is realized by combining their measurements with CBL/VDL measurements. The computer blends the data from both tools to differentiate between gas-cut cement, channeled cement, heavy mud in the casing formation annulus, strong cement, fast formation signals, microannulus, or thin cement sheaths. The calculations are made at the wellsite as an interpretation log—a cement image that color-codes the various possibilities. The casing diameter and thickness are included as well as a GR/CCL log for correlation and VDL waveforms to evaluate the formation bond. And the tools can be run in combination on one trip in the hole.

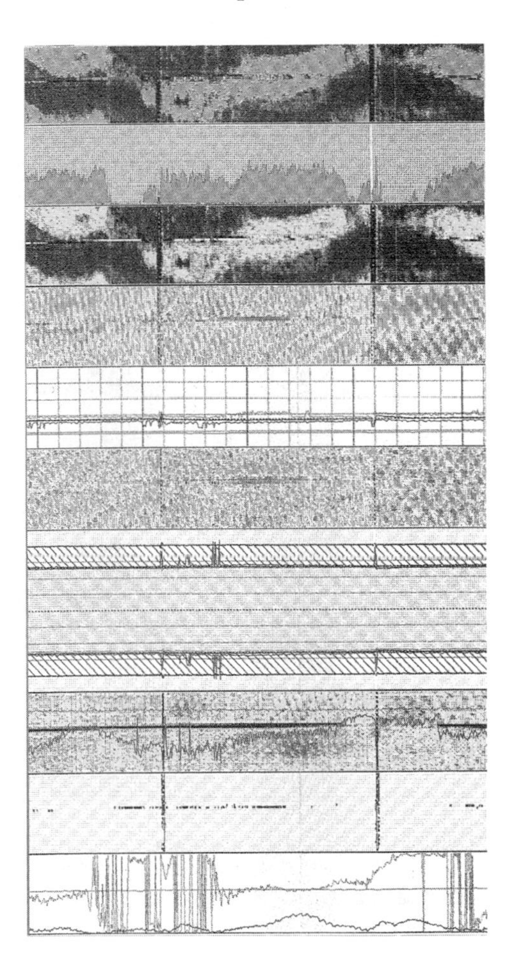

Fig. 11–7. *An UltraSonic Imager gives full 360° coverage. Note the channel shown by the cement map (courtesy Schlumberger).*

Figure 11–8 shows a combination CBL/MSG/CAST/V tool. Track 1 records tool eccentering, tool rotation, and casing ovality as well as the average, minimum, and maximum casing thickness. Track 2 is the CBL amplitude curve, track 3 contains the MSG, track 4 is a CBL BI curve and the average cement impedance determined from the ultrasonic measurements, and track 5 is a color-coded cement map with black or dark brown showing strong cement and white or light blue showing water-filled spaces. In this example, note the channel indicated from X125–X145 ft.

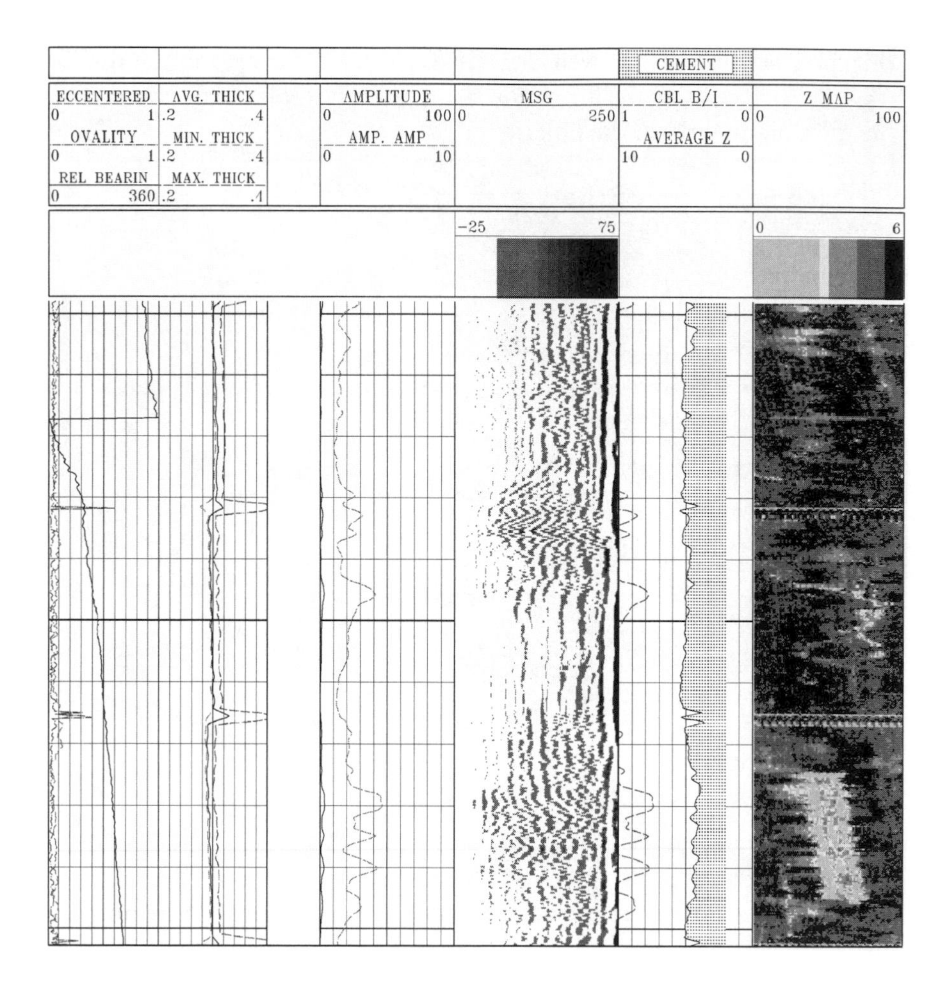

Fig. 11–8. *A combination CBL/VDL and CAST-V log (courtesy Halliburton). By combining the information from the two logs, conditions such as gas-cut cement and heavy mud behind the casing can be identified.*

REMEDIAL CEMENTING

Some type of remedial cement job is usually required when the cement is not providing zone isolation. The definition of poor cement is flexible. A high-pressure gas zone needs better cement than a partially depleted oil zone, for example. A formation that is going to be treated by hydraulic fracturing also needs excellent cement to keep the treatment in zone. Cement evaluation logs help determine whether to try to improve the cement job.

A remedial cementing operation is called a *squeeze job*. The process for most squeeze jobs is similar. Drillpipe or a tubing work string is run in the hole with a retrievable packer set just above the zone to be squeezed. The casing is perforated with four to eight holes, either before the packer is run into the hole or afterward with a perforating gun. If possible, the perforations are *washed* or allowed to flow into the wellbore to clean the debris from the perforation. Next, a pump-in pressure is determined. Cement is then pumped down the tubing and through the perforations. Where it goes from there determines how well the squeeze job worked.

The simplest squeeze job is one in which the cement top was not high enough. The casing is perforated just above the present cement top; then an attempt is made to establish circulation to the surface. If circulation can be established, a calculated volume of cement is pumped down with a wiper plug separating the cement from the displacement fluid. This type of squeeze stands an excellent chance of success.

If circulation is not established, the preceding fluid has to go somewhere so the cement can be pumped in. The fluid fractures the weakest formation and is pumped away, making room for the cement. The cement follows the fluid into the formation. The increase in pump pressure indicates when to stop pumping. The packer is released, and the cement is pumped out. If the formation that breaks down is some distance above the zone of interest, then a successful squeeze may result. However, if the formation that breaks down is a productive zone, then the squeeze may greatly damage the formation.

The most common type of squeeze is called a *block squeeze*. In the block squeeze, the zone itself has poor cement but often there is fair to good cement above and/or below the zone. Figure 11–9 is an excellent example of a block squeeze. After the primary cement job, zones A and B were not adequately protected. It was decided to block squeeze the two zones. The Repair Squeeze section (right) shows an almost perfect repair, but not all squeeze jobs are this successful.

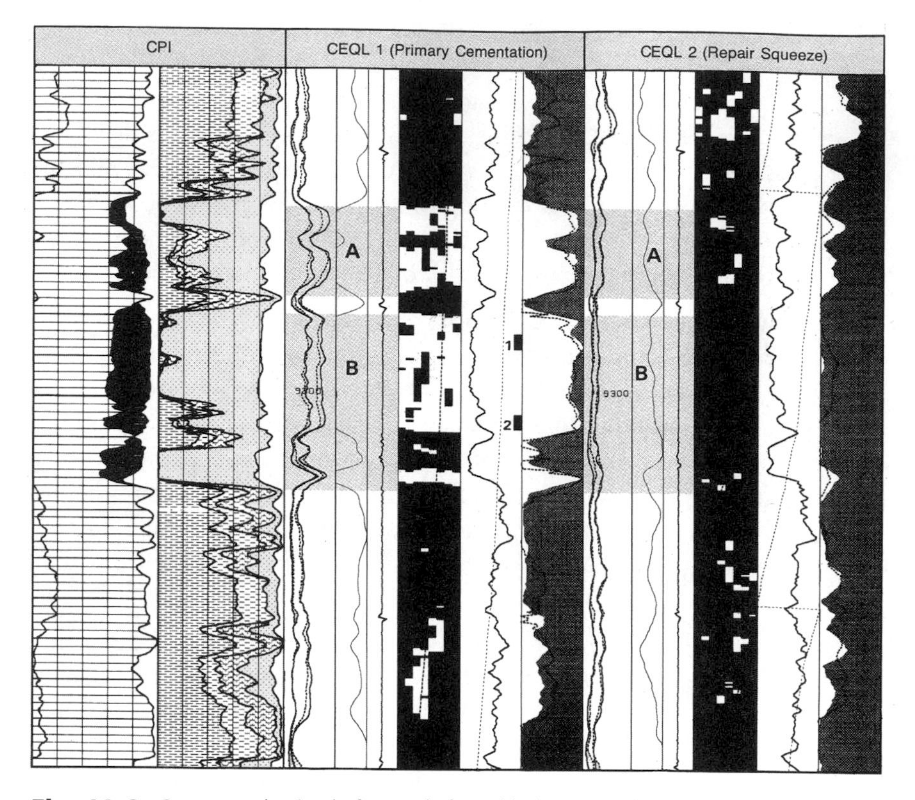

Fig. 11–9. *Cement evaluation before and after a block squeeze (courtesy Schlumberger). Zones A and B show no bonding before the squeeze and almost perfect bonding afterward. Not all remedial cementing jobs are this successful.*

To perform a block squeeze, the area is perforated as above; then about 50 sacks of cement are forced through the perforations. Usually the pumping pressure exceeds the formation breakdown pressure, and the cement is forced into the formation. This often cements any natural fractures and adversely affects the formation's permeability.

In the case of channeled cement, a low-pressure, low-volume squeeze may be attempted. Perforations are made as noted previously. If possible, the perforations should be flowed. The preceding fluid is then pumped at a pressure below formation fracturing pressure, forcing the fluid in the channel back into the formation. The preceding fluid is followed by the cement. The pressure increases suddenly when the cement has filled the channel. The pumps are shut down, but the pressure is held on the well and is allowed to bleed off slowly. Usually only a couple of sacks of cement are needed. The technique is one of finesse rather than brute force, but the results can be excellent.

In the case of gas-cut cement, not much can be done. The gas-cut cement allows gas to migrate up the casing, but it does not allow the injection of liquid into the annular space between the casing and the formation. As in most problems with cement jobs, prevention is the best solution. Additives that prevent gas infiltration of the cement are available and should be used whenever high-pressure gas zones are present in the well.

Prevention is the best solution to all cement problems.

- A careful design and meticulous implementation of that design will avoid most remedial cement operations.
- The wellbore should always be properly conditioned.
- Centralizers and scratchers should be used near zones of interest.
- The casing should be reciprocated (moved up and down) during the cement job.
- Pump rate should be kept to the designed speeds so the flow regime is as designed.

If these factors are followed, then there will rarely be the need for a squeeze job.

PERFORATING FOR PRODUCTION

The only log made while perforating is a GR or a CCL. While production perforating is beyond the scope of this book, the subject is covered briefly to familiarize you with some of the terms. Perforating is obviously important because it lets the oil or gas into the wellbore so it can start on its way to market.

Almost all perforations today are made by shaped charge explosives. The story goes that the technique was discovered inadvertently by someone making gun cotton. The coroner noted the letters *US* embossed on the discoverer's forehead and surmised it was somehow transferred from the package in which the cotton had been contained. The next advance was when a different coroner noted copper traces in the head wound of a person who had been making dimples in blasting caps. However, it took WWII to advance the science sufficiently to make holes in steel. The technique was used in bazooka rounds to penetrate tank armor. After the war, the technology was adapted by Welex to perforate oilwell casings. Although the perforating charges have been greatly improved, the theory is still the same.

Figure 11–10 shows various factors considered when designing a new perforating charge. The shaped charge consists of a shell or housing, a cone-shaped explosive, and a liner. The mass of the housing gives inertia to the charge so that most of the energy is directed opposite to the housing. The explosive, of course, supplies the energy. The angle of the cone is critical to the performance of the charge. The liner is usually copper in addition to some other heavy metal, such as zinc or titanium. The copper is volatilized by the heat of the explosion. The weight of the copper gives mass to the jet of hot gases created when the charge goes off. The jet of gases and volatilized metals blows through the steel of the casing, leaving a smooth, round hole. The cement and formation are also penetrated by the shaped charge jet, aiding in production of the formation fluids.

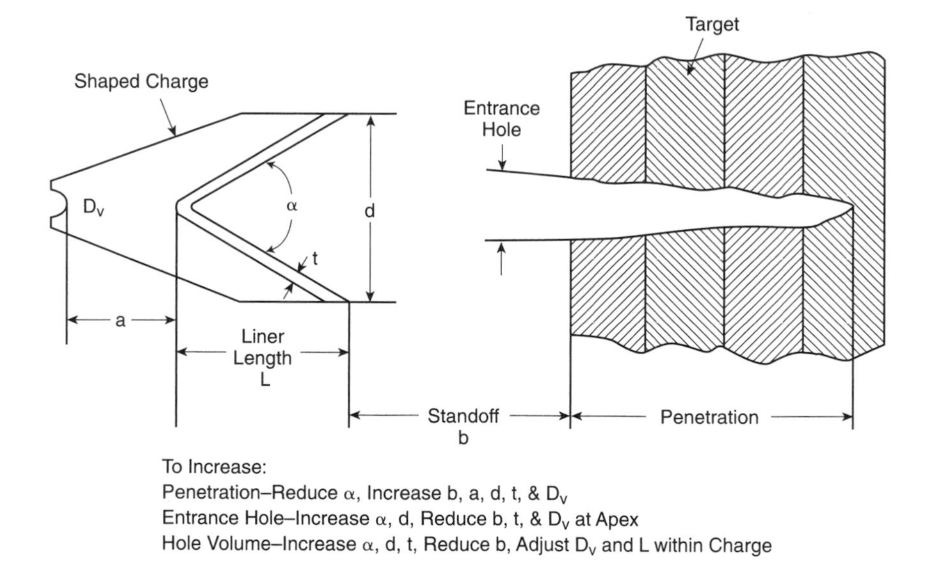

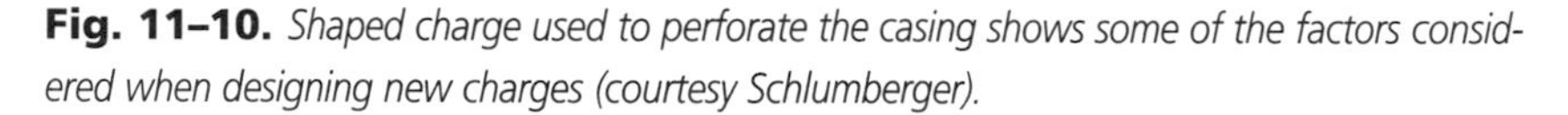

Fig. 11–10. *Shaped charge used to perforate the casing shows some of the factors considered when designing new charges (courtesy Schlumberger).*

The charges are usually contained in a hollow steel carrier called a perforating gun. The explosion is initiated by an electrically ignited blasting cap. The cap sets off a string of primacord that connects all the charges. The logging company personnel who handle these guns are highly trained and very safety conscious. Multiple safety devices must be used, and redundant safety rules must be followed before any explosive device is attached to the wireline.

There are three main types of perforating completions (see Fig. 11–11): casing guns on wireline with hydrostatic pressure greater than reservoir pressure; through-tubing guns on wireline with hydrostatic pressure less than reservoir pressure; and tubing-conveyed guns with hydrostatic pressure less than reservoir pressure.

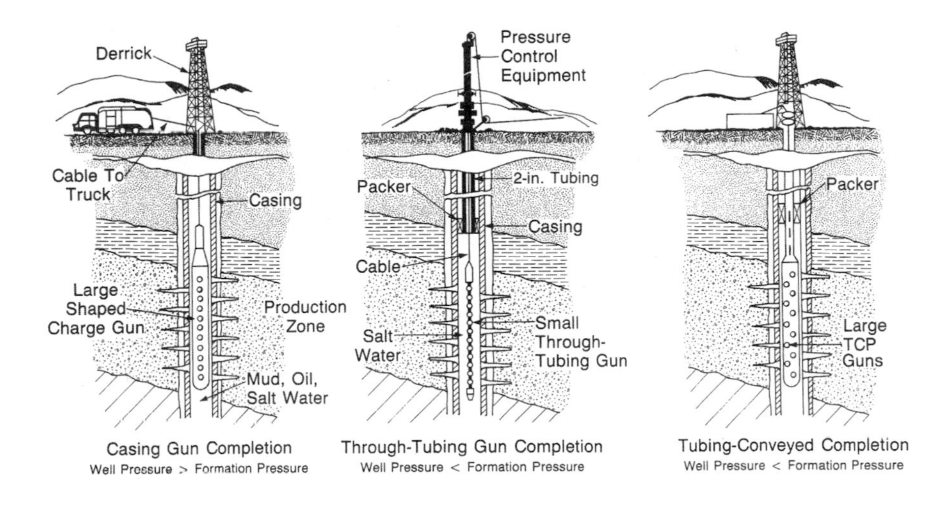

Fig. 11–11. *The three types of perforated completions (courtesy Schlumberger).*

Casing guns perforate the casing before tubing is run in the hole. The casing gun has the largest size of charge that can be used in a given casing size. For example, a 5-in. casing gun can be used in 7⅝-in. casing. There isn't space to safely run a larger gun. Other casing gun sizes commonly available are 4 in. and 3⅜ in. These sizes allow the use of a larger explosive charge with more power. The number of shots, or holes, per foot is adjustable up to four shots per foot, and the shots are usually phased at 60°–90° to each other.

Through-tubing guns are sized so they will fit through the ID of the commonly used production tubing, 2½- and 2-in. ID. The guns are either 2⅛- or 1¹¹⁄₁₆-in. OD. The charges are obviously much smaller with much less penetration power. The shots per foot can be up to 4 ft., the same as the casing guns. The shots in through-tubing guns are all in the same plane, called *zero phasing.*

Tubing-conveyed guns are essentially casing guns run into the hole on tubing. They are used in special conditions such as very-high-angle holes (up to horizontal) or when very long intervals must be perforated and the rig time is very expensive. These conditions are generally on offshore rigs.

One safety note is in order. Any time perforations are made in the casing, the result will be formation pressure connected to the surface. It is imperative that wireline pressure control equipment be used whenever perforations are open.

The well has been drilled and logged, casing has been run and cemented, completion logs have been made, and the well has been perforated. Now we can sit back and watch the money roll in. Right? Well, maybe. In the next chapter, we'll look at some of the problems that can and often do arise in a production or injection well. Fortunately, many of these problems or conditions can be discovered, analyzed, and sometimes even solved with wireline logs and tools.

12

Monitoring the Well and Reservoir

In this chapter, we'll look at the producing well (and the injection well) and some of the problems that can be evaluated and/or remedied by wireline logging tools. Many of these problems are concerned with the tubular goods, casing and tubing; other problems have to do with the reservoirs themselves. We'll look at some new technology in evaluating flow problems in highly deviated wells and in monitoring fields undergoing waterfloods or other secondary and tertiary recovery projects. As oil prices rise and reserves shrink, esoteric methods of increasing production become more feasible. Fortunately, the logging industry is keeping pace with innovative techniques to handle oil companies' demands for better information.

PRODUCTION LOGS

There are only two types of wells: producing wells and injection wells. *Producing wells* are the ones that make the money; the oil or gas flows to the tank or pipeline. Injection wells are normally used to inject fluids into a reservoir for secondary or tertiary recovery projects. Production logs (PL) can be run on either type of well, normally under dynamic conditions (while production or injection is taking place).

Production log is a catch-all phrase for several measurements that may help evaluate production or injection problems. It is important to know where the produced fluids are coming from and where the injected fluids are going. Production logs can determine this information.

Producing wells are the most common candidates for production logs. In a complicated, expensive completion, it may be useful to run a series of production logs shortly after the well is completed. These logs show which zones are producing, how much each zone is producing, and whether the production is oil, gas, water, or a mixture of the three. This is similar to having a baseline EKG in your medical file. Such a record can be very useful in showing whether the well is producing as expected or if there are mechanical problems. However, the normal case is to wait for problems to show up before running production logs.

COMMON WELL PROBLEMS

The production engineer compares the initial results with the expected results based on the log information, and possibly data from nearby wells producing from the same formation. If the well performs as predicted, all is good. If the well is not living up to expectations, then further investigation is warranted. Sudden changes in production, such as a rapid increase in either water or gas production in an oil well, a rapid decrease in production, or increasing pressure in the casing–tubing annulus, indicate downhole problems that should be investigated.

An investigation is usually begun by the production engineer in charge of the well. First, the problem is defined. Generally, the problem is either no production, less production than expected, or different fluid production (water or gas) than expected. Second, all well information is reviewed—the drilling data, mud logger's records, open-hole logs, cement evaluation logs, perforating records, test results, and any other data thought to pertain to the problem. Third, any additional information, the methods available to gather the information, and the elements involved in obtaining the information, are weighed. Decision-makers must be satisfied that the production logs give enough information to identify the cause of the problem. In addition, a cost-effective, relatively low-risk solution to the anticipated problem must exist. Once all the information has been assessed, a decision and an operation plan are made.

Before we discuss production logging measurements and how they are used, let's look at some of the problems that wells commonly experience.

Crossflow

A very common well problem is crossflow, when two formations at different reservoir pressures are in communication. Instead of flowing to the surface, the production is stolen by a thief zone. Crossflow may occur inside the casing between two or more perforated intervals or between perforations and a casing or packer that is leaking to a low-pressure zone. If the flow is outside the casing, it is from channeled cement or natural fractures in the formation.

Channeled cement allows unwanted fluids such as water or gas to enter the production perforations. In an injection well, the channel allows the water to charge the wrong formation. Channels may extend for hundreds of feet and allow communication with formations that were thought to be too far away to be a problem.

Premature Breakthrough

A serious problem in waterfloods is premature breakthrough of the flood front. If multiple zones are open or if the interval is thick (15 ft. or more), variations in permeability may allow the flood to advance more quickly through one zone or portion of a zone. Once the water breaks through, it quickly chokes off the oil production, and water production increases dramatically. Recognizing this problem early allows countermeasures to be taken.

Mechanical Problems

A variety of mechanical problems may occur in both producing and injecting wells (Fig. 12–1). Casing leaks due to corrosion can allow unwanted water entry. Packers may leak, allowing pressure into the casing tubing annulus. If there is a hole in the casing above the packer (maybe from a previously perforated zone that was unproductive), a thief zone may be created. Water may leak around faulty plugs. Perforations may have been made out of zone—slightly too high or too low or in the wrong place completely. Perhaps the gun misfired and no perforations were made at all. It happens.

Underproduction

A less obvious problem is underproduction. Zones that appear productive on the open-hole logs may produce far below expectation. The permeability may have been damaged during drilling or completion. Often, a hydraulic treatment, such as acidizing or fracturing, is all that is needed to let the well produce up to expectation.

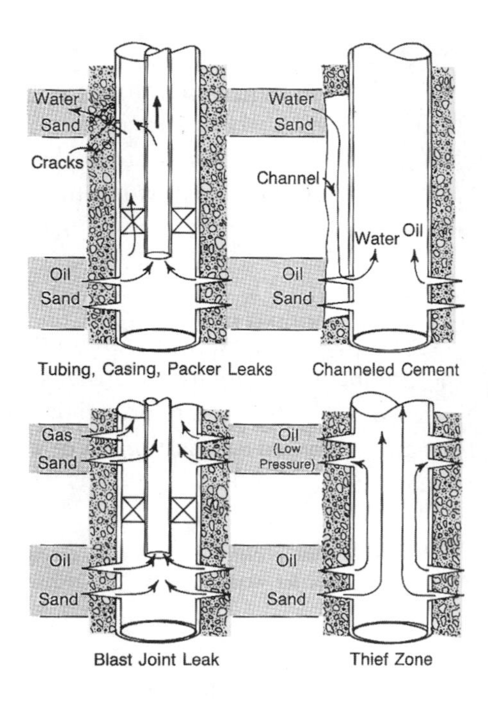

Fig. 12–1. *Common mechanical problems encountered in both production and injection wells (courtesy Schlumberger).*

The information needed to identify any or all of these problems is the same:

- Where are the points or intervals of fluid entry/exit?
- What is the flow rate at each point of fluid entry/exit?
- What is the type, or mix, of fluids at each entry point?
- What is the origin of the fluids coming into the well at each interval?

Production logging tries to answer the question, "How *much* of *what* is coming from *where*?" It sounds simple, but it's not. Let's see what complicates the procedure and makes production analysis a lot closer to rocket science than it is to plumbing.

THE COMPLEXITIES OF PRODUCTION ANALYSIS

In a flowing water well, the solution to our production logging question is straightforward. The production at the surface can be measured in gallons per hour or barrels per day; the water can be seen, tasted, and checked for salinity. The only unanswered question is, "Where is the water entering the well?" A flowmeter log determines where the water is entering and compares the downhole flow rate to the surface flow rate. A temperature log and pressure gauge are normally included in the logging toolstring. The temperature log helps verify the water is coming from the perforated formation and not from elsewhere. If the origin of the water is in doubt, a water flow analysis can be made with a pulsed neutron tool (discussed later in this chapter).

Unfortunately, evaluating oil and gas production is less straightforward. The produced fluids are very complex, and the downhole flow of these fluids is even more complex. Let's consider an oil reservoir at its original conditions—that is, the pressure and temperature are the same as they were when the oil field was discovered. Under these conditions, each barrel of oil contains some dissolved gas.

Let's take a barrel and change the pressure and temperature from reservoir values to standard surface conditions, 14.7 psi and 70°F. As the pressure and temperature decrease, the dissolved gas begins to come out of solution and the oil volume contracts. The change in oil volume is called the *shrinkage factor* (B_g). The gas volume varies directly with pressure and inversely with temperature according to Boyle's gas laws. The gas that started out as a small fraction of a barrel at reservoir conditions increases in volume to perhaps 200 bbl. Gas is measured in cubic feet (cfg) instead of barrels. The relationship between surface gas and surface barrels of oil is called gas/oil ratio, or GOR. The value of GOR can range from 0–20,000 cfg/bbl or more for dissolved gas.

The amount of solution gas can be determined by taking a fluid sample at downhole conditions. A PVT (pressure/volume/temperature) tool recovers a fluid sample in a small chamber and seals the sample at recovery conditions close to reservoir conditions. (The well is usually shut in for this process so the pressure and temperature are as close to static reservoir conditions as possible.) The sample is then transported to a laboratory where the characteristics of the recovered oil and gas sample are measured. This process gives an accurate relationship of GOR, oil gravity, oil volume shrinkage factor, oil and gas chemical composition, reservoir bubble-point pressure, and other data of interest to the reservoir engineer. If a PVT sample has not been taken, an analysis of the oil and gas may be made from samples collected at the surface. While not as accurate as PVT data, API chart values based on the surface sample analysis can be used for oil density, GOR, and gas density at downhole flowing conditions.

Multiphase Flow

A gas well making no water has only a single-phase fluid just as the water well discussed earlier. If the gas well also produces water, a two-phase fluid is flowing. In an oil well with both gas and water production, the well has three-phase fluid flow, and the situation becomes even more complex. When more than one fluid phase is present, the *holdup* of each phase must be determined. Holdup, y, is the volumetric percentage of each phase. Therefore, the sum of the individual holdups must equal 1.0, or 100% ($1 = y_G + y_O + y_W$).

When fluids of different densities flow in the same well, the lighter phase flows faster than the heavier phase. The difference in velocity caused by the difference in density of the two fluids is called *slip velocity*. Empirical relationships for slip velocity have been used for years. But recent research for very-high-angle drilling (up to horizontal) technology has provided new understanding in determining slip velocity as well as all parameters of multiphase flow.

To accurately measure downhole flow rates of the three fluids, the velocity and quantity of each phase must be measured. To understand how the three phases change characteristics as the amount and velocity of each phase changes, we must first understand the various flow regimes.

Flow regime is the term used to describe the various flow patterns that occur at different flow rates by multiphase fluids (see Fig. 12–2). Flow regime is determined mainly by the relative flow rate of each phase. Water and oil at low flow rates are in *laminar flow*; at high flow rates the two liquids are in *turbulent flow*. If gas is flowing with a liquid, the decrease in pressure as the fluids rise up the casing allows the gas to expand and continuously change its holdup, y_G. If oil with solution gas is flowing from a well, the flow starts out as a single-phase fluid, oil. As soon as the flowing pressure drops below the bubble-point pressure, the gas begins to come out of solution as small bubbles and $y_O \approx 1$ and $y_G > 0$. At this point we have *bubble flow*. As the pressure decreases more, the gas expands and we get *plug* or *slug flow*. The holdups have changed, so now $y_O \approx 0.4–0.5$ and $y_G \approx 0.5–0.6$. As the pressure continues to decrease and the gas continues to expand, the flow goes through froth flow to mist flow with $y_O < 0$ and $y_G \approx 1$.

The same flow regimes can exist with water and gas. The total flow rate affects the flow regime also. If the flow rate is high or the reservoir pressure is below bubble-point pressure, the flow regime might start with slug flow and advance quickly to mist flow. On the other hand, if the flow rates and surface gas production are low, then the flow may stay in the bubble or plug flow stage. With three-phase flow, the various flow regimes are similar but are still not completely understood.

A further complicating factor in three-phase flow is hole deviation. The high hole angles affect the flow regimes by the added factor of *gravity segre-*

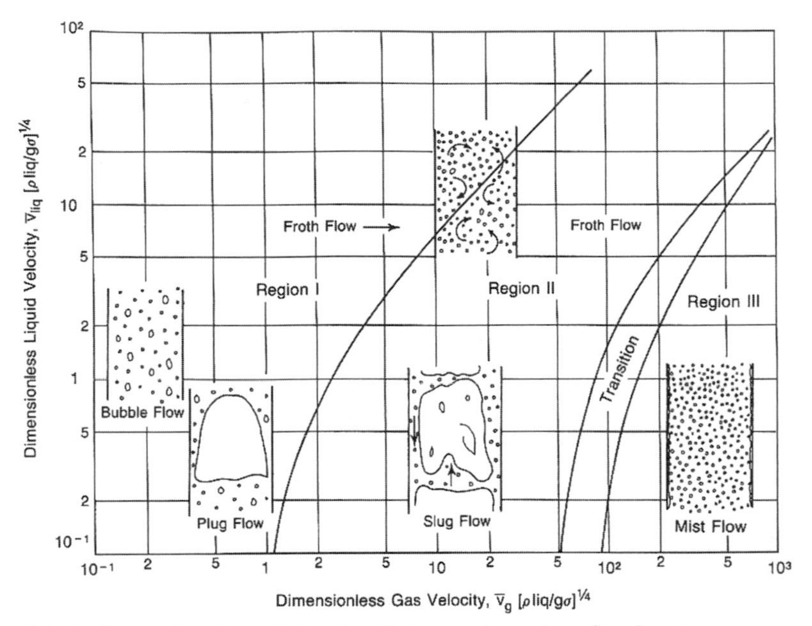

Fig 12–2. *Flow regimes in vertical wells with two- or three-phase flow (courtesy Schlumberger). Changes in flow regime are caused by hole angle, relative flow rates, and both differences and changes in fluid density.*

gation. In a horizontal well with three-phase flow, theoretically the flow stratifies, with gas flowing in a layer at the top of the casing, oil in the middle stratum, and water along the bottom of the casing. In actuality, turbulence often causes mixing and rapid changes in holdups. Often the oil or water is dispersed as drops in the more abundant liquid (oil drops in water or water drops in oil). The liquids tend to be on the low side of the casing and the gas is on the high side.

PRODUCTION LOGGING TOOLS

The production logging tools of the 1980s could not measure flow rates accurately at very high deviations because of segregation effects. Additionally, these older tools measured the average flow across the casing rather than measuring points or small areas. With the conventional tools of just a few years ago, only two-phase flow could be accurately determined. In cases of three-phase flow, holdup was divided into gas holdup and liquid holdup. The slip velocity between oil and water was assumed or neglected, and an average density was calculated based on the assumed water and oil holdups.

Research carried out during the last few years has yielded better measurements in difficult flow conditions. New techniques and logging devices (Table 12–1) have resulted in better answers to the question, "How much of *what* from *where*?"

The production logging equipment is sized to fit through commonly used production tubing. Most of the devices are 1 11⁄16 in. in diameter to fit in 2-in. tubing.

Production Logging Tools	Measures
Gamma ray	Radioactivity; correlates to open-hole logs; RA tracer surveys
Collar locator	Casing collars for depth control
Temperature	Wellbore temperature changes due to production
Flowmeter	Downhole flow rate
Manometer	Pressure and differential pressure
Differential manometer	Average density of the fluid column
Holdup meter	Percent, by volume, of the fluid column that is either gas, oil, or water
Pulsed neutron	Water flow inside or outside the casing
Radioactive tracer	Radioactive fluid movement with the gamma-ray tool
Casing Inspection Tools	
Casing potential tool	Changes in voltage from galvanic action
Acoustic imaging tool	Casing ID and OD; indicates pitting and corrosion
Magnetic flux leakage	Changes in magnetic flux due to casing anomalies such as pits and holes
Multifinger caliper	Changes in casing ID at many radii (16–80)
Through-Casing Evaluation Tools	
GR spectrometry	Percent of natural GR radiation from thorium, uranium, and potassium
CNL	Neutron porosity
Acoustic	Acoustic porosity; need well-bonded casing
Density	Density porosity; need exceptionally well-bonded casing; seldom possible
Pulsed neutron spectrometry	Elemental yields of carbon, oxygen, silicon, calcium, iron, and sulfur; determine water saturation and lithology
Pulsed neutron decay rate	Sigma; determine water saturation and porosity

Table 12–1. *Cased hole logging tools and what they measure*

GR and CCL Tools

The gamma-ray log and the casing collar locator are used for depth control to correlate to previously run open- or cased-hole logs. In a well that has been on production for some time, usually years, *hot spots*—areas of increased GR readings when compared to the original logs—are sometimes seen. These hot spots are from radioactive salts, dissolved in the formation water, that accumulate on the casing. As such, the hot spots indicate water entry. The GR tool is also used with the radioactive (RA) tracer ejector tool.

Temperature Tools

The temperature tool determines fluid entries—especially gas. When gas leaves the reservoir and flows through a perforation, pressure drops and the gas expands. The gas temperature decreases because of adiabatic expansion.

One way to determine flow behind the casing is to shut in the well for several days so the temperature can stabilize. A base temperature log is then run, and the well is put back on production. A series of temperature, flowmeter, fluid density, and pressure logs are made as the well flows. The change in temperature from the base log shows whether fluid is flowing either inside or outside the casing; the flowmeter and the fluid density logs show where the fluid is entering the well.

Flowmeters

Flowmeters use several different techniques to measure downhole flow rates. A four- to six-bladed propeller spins, or rotates, as fluid moves across the propeller blades. The tool counts the revolutions per second (rps) of the propeller. The speed of rotation is proportional to flow rate.

Several factors must be kept in mind when interpreting a flowmeter.

- The spinner measures the flow relative to the tool, so the speed at which the tool moves up- or downhole must be factored in. For this reason, tool speed is recorded on the log.

- The cross-sectional area of the casing must be known in order to calculate flow rate from the spinner. Usually, the casing ID is taken from published tables. Better results can be obtained if the casing ID is measured rather than assumed. Some of the newer flowmeter tools include an *x-y* caliper—two measurements of ID at right angles to each other.

- The spinner tools must be centralized since the flow velocity inside the casing is not uniform across the width of the casing. Depending on the flow regime, various flow correction factors account for the variations in flow profile. The corrections to flow velocity are probably not accurate in deviated wells.

Full-bore spinners (FBS) are one variation of flowmeters. The FBS collapses to a small diameter while running through the tubing. Once the tool exits the tubing, the caliper/centralizer arms and the spinner blades open up to a larger diameter. The caliper and centralizers hold the tool in the center of the casing. The spinner blades are selected based on the casing size, using the largest diameter spinner assembly possible. The four- to six-arm caliper acts as a centralizer, protects the spinner blades from damage, and records an *x-y* caliper curve.

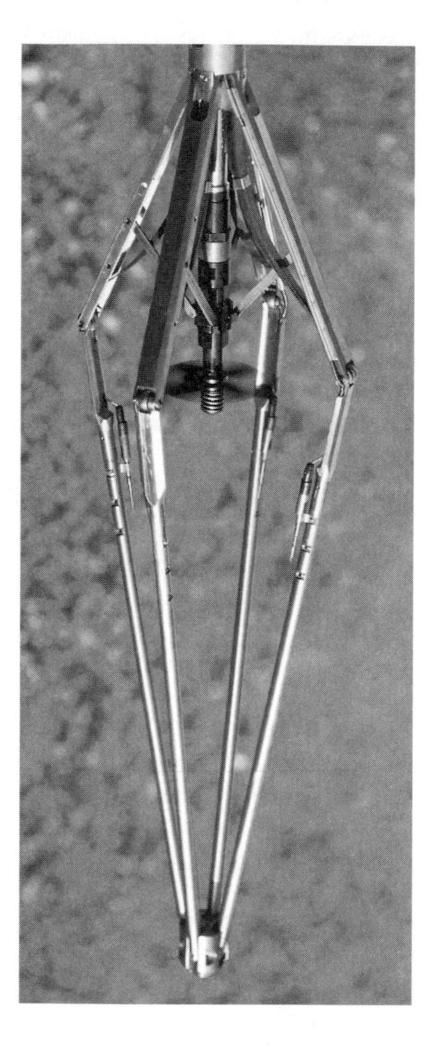

Fig. 12–3. *Full-bore spinner tool with x-y caliper and water holdup probes mounted on the arms (courtesy Schlumberger). The tool measures flow rate and casing diameter in two directions as well as holdup at four points in the casing.*

On some of the newer models (Fig. 12–3), electric probes mounted on the caliper arms measure water holdup and make a bubble count at four points in the same plane as the spinner. These holdup measurements show the distribution of fluids within the casing as the flow rate is measured—very important in rapidly changing flow regimes or in high-deviation wells where the gravity difference of the produced fluids (especially gas) greatly affects the flow pattern. The *x-y* caliper allows an exact calculation of cross-sectional area to help determine flow and casing damage.

The *diverter flowmeter* is yet another spinner tool (Fig. 12–4). It uses an umbrella-like skirt around a centralizer to divert the flow through the spinner blades. The tool is run through the tubing with the diverter umbrella closed. As the tool exits the tubing, it is positioned above the first set of perforations. The diverter is opened, making a positive seal with the casing wall so none of the fluid leaks past, and takes stationary measurements at the selected interval. Then the diverter is closed, the tool is moved to a new position, and the diverter is reopened. The process continues until the tool is below all fluid entries.

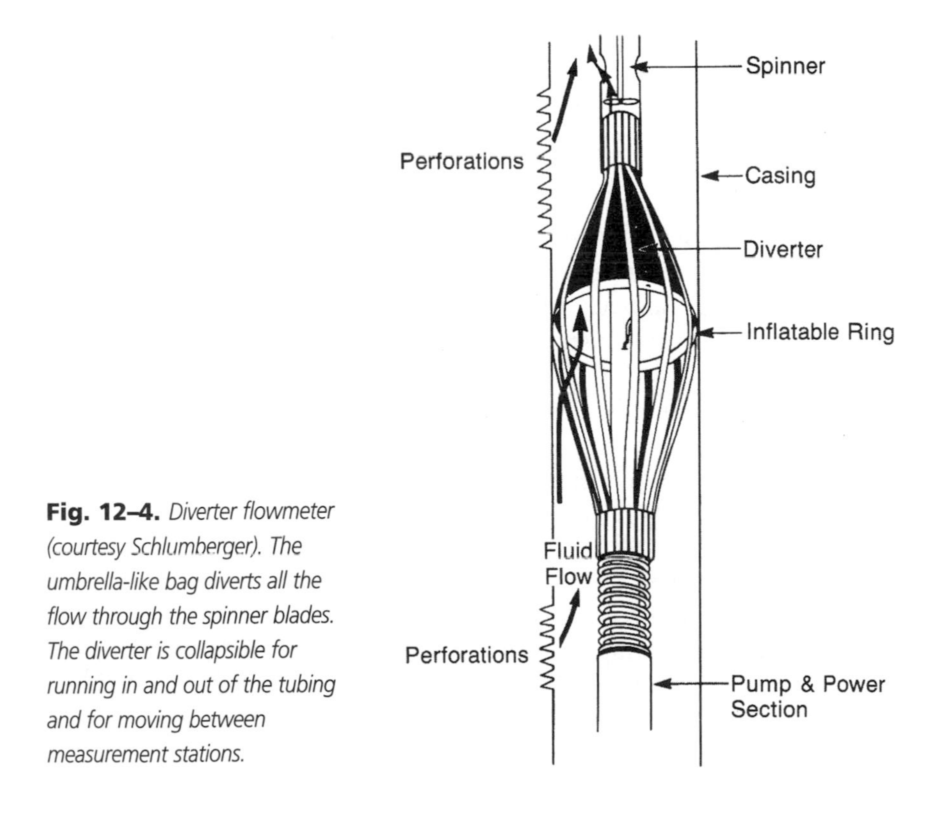

Fig. 12–4. *Diverter flowmeter (courtesy Schlumberger). The umbrella-like bag diverts all the flow through the spinner blades. The diverter is collapsible for running in and out of the tubing and for moving between measurement stations.*

Still another device is the *continuous spinner*, which also uses a rotating propeller to measure flow rates. This spinner, however, has a fixed diameter small enough to fit inside 2-in. tubing. The tool is normally used to measure flow in tubing where the velocities are high and the produced fluid is homogenized by the high flow rate.

All spinner devices determine flow rate the same. The change in rps is related to changes in downhole flow rates. In most cases, the spinner output is calibrated in flow rates of b/d and/or cfg/d. Sometimes the output is presented as a percent of total flow, such as zone 1 = 10%, zone 2 = 35%, zone 3 = 0%, and zone 4 = 55%. In an injection well where the total fluid being pumped is accurately known and the fluid is a single phase, this result is adequate.

Spinner and tracer tools were the only way to measure downhole flow rates until recently. One of the newest flowmeters is the Multi-Capacitance Flow Meter (MCFM) developed by Baker-Atlas and Shell International E&P (Fig. 12–5). The MCFM uses 28 sensors on a wing that extends from the centralized tool to both sides of the casing. The wing is oriented vertically in high-deviation wells. The sensors determine percentages of gas, oil, and water, and the flow rates of each. The obvious advantage is that complete holdup information as well as individual flow rates are obtained simultaneously from either continuous or stationary readings. The tool, developed for pipeline use, works very well in horizontal wells.

Manometers

Reservoir pressure, flowing pressure, and pressure gradients all yield useful information to the petroleum engineer. The manometer is a very accurate pressure gauge normally run in the production logging toolstring. The pressure data it collects is used with the temperature to calculate the downhole fluid density of each fluid present. Gauges also obtain pressure buildup data by shutting in production for an extended time. When only pressure buildup data is wanted, the pressure gauge is run on *slick line* (small-diameter wireline that has no electrical conductors) with a battery-operated recorder. The assembly is left for several days and then retrieved.

Differential Manometer

Fluid density (ρ_F) is one of the most important parameters measured—needed to calculate holdup and slip velocities. However, fluid density measurements can only determine two-phase holdups since the applicable equa-

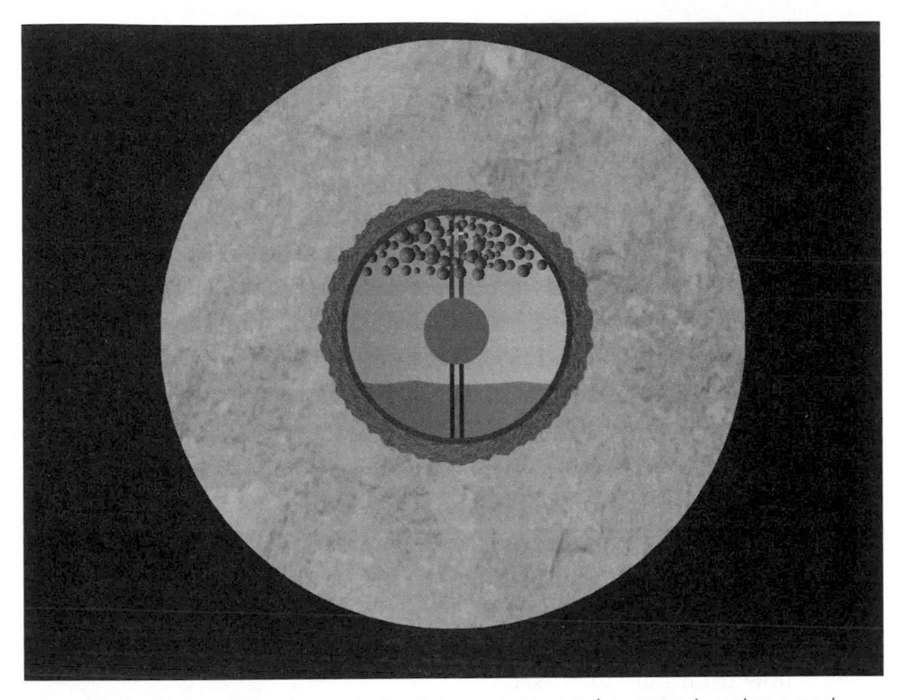

Fig. 12–5. *Multicapacitor flowmeter has 28 sensors mounted on two wings that extend across the casing (courtesy Baker Atlas). The tool can measure both a flow and holdup profile in both vertical and high-angle wells. The tool was developed in conjunction with Shell International E&P.*

tions are indeterminate with three unknowns but can be solved for two unknowns. For two-phase holdup,

$$1.0 = y_H + y_L$$
$$\rho_F = \rho_H y_H + \rho_L y_L$$

where:

H = heavy phase
L = light phase

For three-phase holdup:

$$1.0 = y_W + y_O + y_G$$
$$\rho_F = \rho_W y_W + \rho_O y_O + \rho_G y_G$$

The density of the water, oil, and gas are determined from handbook values or PVT measurements in conjunction with the measured values of pressure and temperature.

The ρ_F measurement is made in several different ways. The simplest scheme is to measure the change in pressure with depth on the manometer pressure gauge, divide by the change in depth, and convert to the proper units. Another method is to use the difference in hydrostatic pressure on two very sensitive pressure gauges (or pressure bellows, in some tools) separated by a short, fixed distance. The bellows measurement is more sensitive than the pressure gauge's; however, it must be corrected for friction effects at high flow rates. Both of these tools are adversely affected by high hole angles. The vertical distance between the transducers is reduced by the hole angle, thereby reducing the change in hydrostatic pressure. The apparent ρ_F can be corrected if the hole angle is known.

The new approach to determining fluid density adapts open-hole density technology. As with the conventional open-hole density tool, a GR source emits low-energy gamma rays that interact with the fluids in the wellbore. The measurement is the electron density of the fluid mixture in the casing. As we know from our study of the density tool, electron density is proportional to actual density, so the measurement is a volume-weighted average density of the fluids present. The fluid density can be determined in either vertical or highly deviated wells since gravity plays no part in this measurement. The only disadvantage is the variation in value due to the statistical random decay rate of radioactive measurements.

Holdup Meters

Fluid density measurements are the traditional way to determine holdup. New technological developments have resulted in several additional methods that determine the volume of each fluid phase present. Holdup meters, such as electric probes on the full-bore flowmeter or capacitance sensors on the MCFM, measure holdup directly. In addition, these sensors give a distribution pattern of the holdups in different areas of the wellbore.

The probes count the rate and size of the bubbles by measuring changes in the fluid conductance. The holdup distribution and bubble count help determine the flow regime in a particular region of the well. This information is especially important in deviated wells where the difference in gravity of the various fluids combined with gravity segregation due to the hole angle cause very complex flow patterns. The MCFM measures the capacitance at

28 different sensors. By interpreting capacitance changes, both holdup and flow rate of each phase can be determined.

New gas holdup tools have made it possible to calculate three-phase holdups by eliminating the third unknown quantity in the fluid density equations. Halliburton's Gas Holdup tool (GHT) uses a density device to measure gas volume. Schlumberger's Gas Holdup Optical Sensor tool (GHOST) uses fiberoptic technology to detect and count bubbles in a liquid stream or liquid droplets in a gas stream.

Pulsed Neutron Tools

Although pulsed neutron tools were designed primarily for reservoir monitoring through casing, they have a role in evaluating production problems. The pulse of neutrons activates the nuclei of oxygen atoms. An activated oxygen atom in turn releases a characteristic gamma ray as the atom returns to a neutral state. The energy level of gamma rays from oxygen activation is detected by the tool/detector circuitry. The oxygen is primarily in the formation and borehole water, so the detection circuitry is adjusted to be influenced mainly by the water in or near the wellbore. Several strategically located detectors measure the response to the activated oxygen.

The tool can determine water flowing up or down and either inside or outside of the casing. It is also useful in detecting the origin of unwanted water and locating water flowing through channels in the cement. And it can measure water holdup.

Radioactive Tracer

The radioactive tracer ejector tool (RA tracer) is old technology that is still useful in some situations. Tracer surveys can often locate channels and casing/packer leaks that cannot be identified with certainty by other methods. In addition, the tool can measure flow rates between zones. The RA tracer is used mainly in injection wells but can also be used in producing wells. Since the small amount of ejected radioactive material is mixed into a large volume of other fluids, there is no measurable increase in radiation at the surface. The half-life of the RA fluid is about 45 days. Properly handled, there is no health risk in a tracer survey. However, a pulsed neutron tool in water activation mode can be used in place of the tracer tool if concerns exist about the release or storage of radioactive liquids.

In a typical tracer survey, the toolstring usually includes three GR detectors and the ejector pump. In an injection well, one GR detector is located above the ejector and two detectors are below the ejector. The distances between the ejector pump and the GR detectors are measured and noted on the log heading and are used to calculate flow rates. The recordings are normally made with the tool stationary and the recording on time drive. The radioactive solution is ejected, in measured spurts on command, by a small pump. The path of the fluid is then followed by watching the response of the various GR detectors. Usually 3–10 different ejections and log passes are made to define a problem completely.

Figure 12–6 illustrates a *velocity shot* in an injection well. The RA tracer is stopped at a specific depth, and the recorder is put on time drive. Tracer fluid is ejected, with the time indicated automatically on the log. As the injection continues, the tracer fluid moves down past the two detectors located below the ejector pump. The time it takes for the slug of tracer fluid to pass the detectors, divided by the distance between the detectors, is the flow velocity. By combining the fluid velocity with the flow area, the flow rate can be determined:

$$Q = [6.995\,(D^2 - d^2)S]/t$$

where:

Q = flow rate, b/d

6.995 = conversion constant

D = casing ID, in.

d = tool OD, in.

S = spacing between detectors, in.

t = time for slug to move between the detectors, sec

Normally, the first shot is made above all perforations so the total flow rate can be established. The upper detector (above the ejector) is monitored for a response. If tracer fluid is seen on the upper detector, upward flow is indicated. Since the flow in the casing is down (in an injection well), the counterflow can only come from a channel between the casing and the formation. (Naturally, everything is reversed in a producing well.) Then the tool is moved below the first set of perforations and the process is repeated. If multiple sets of perforations are present, several stations are needed to define the injection profile and to determine any leaks or channels. At the conclusion of the stationary readings, a large slug of radioactive fluid is ejected and then followed with the tool. Several timed passes are made until the tracer fluid is no longer visible on the log. This procedure may be repeated several times, depending on the results obtained.

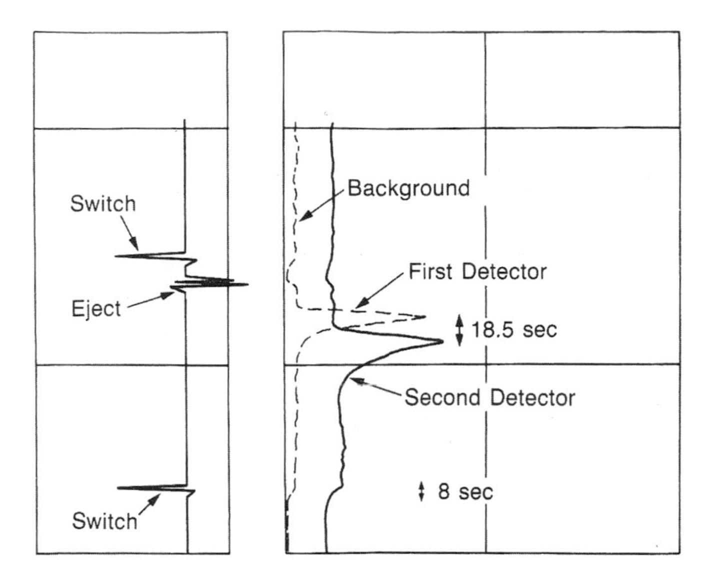

Fig. 12–6. *RA tracer velocity shot (courtesy Schlumberger). By timing the passage of the radioactive fluid between two detectors, the fluid velocity and therefore the flow rate can be calculated.*

In Figure 12–7, the object was to determine if the injected fluid was staying in zone (sands 2 and 3) or possibly channeling into either sand 1 or 4. The tracer fluid was ejected just below the packer, and a single GR log was run several times. On the first run, the slug was still in the casing, as shown by the large response at *a*. The smaller response at *b* was from radioactivity trapped in the turbulence at the packer. As the slug moved down, its path can be followed. The peaks at *a, c, e, h, l,* and *p* record the slug's path as it moved down the casing. Peaks *l* and *p* indicate a casing leak below zone 2 or, more likely, a channel from zone 2 to zone 1. Peaks *f, j, h,* and *v* indicate flow from zone 3 to zone 4 via a channel. Peaks *i, m,* and *q* indicate sand 3 is taking fluid. Sand 2 shows no indication of fluid flowing into it but just into the perforations and down the channel to zone 1.

EXAMPLE PRODUCTION LOG

Figure 12–8 shows a simple production log in a gas well. Water production has steadily increased over the past months. Production has fallen due to the decrease in flowing pressure caused by the water production. Four sets of perforations are open. The operator wants to know where the water is coming from and whether it can be reduced or eliminated.

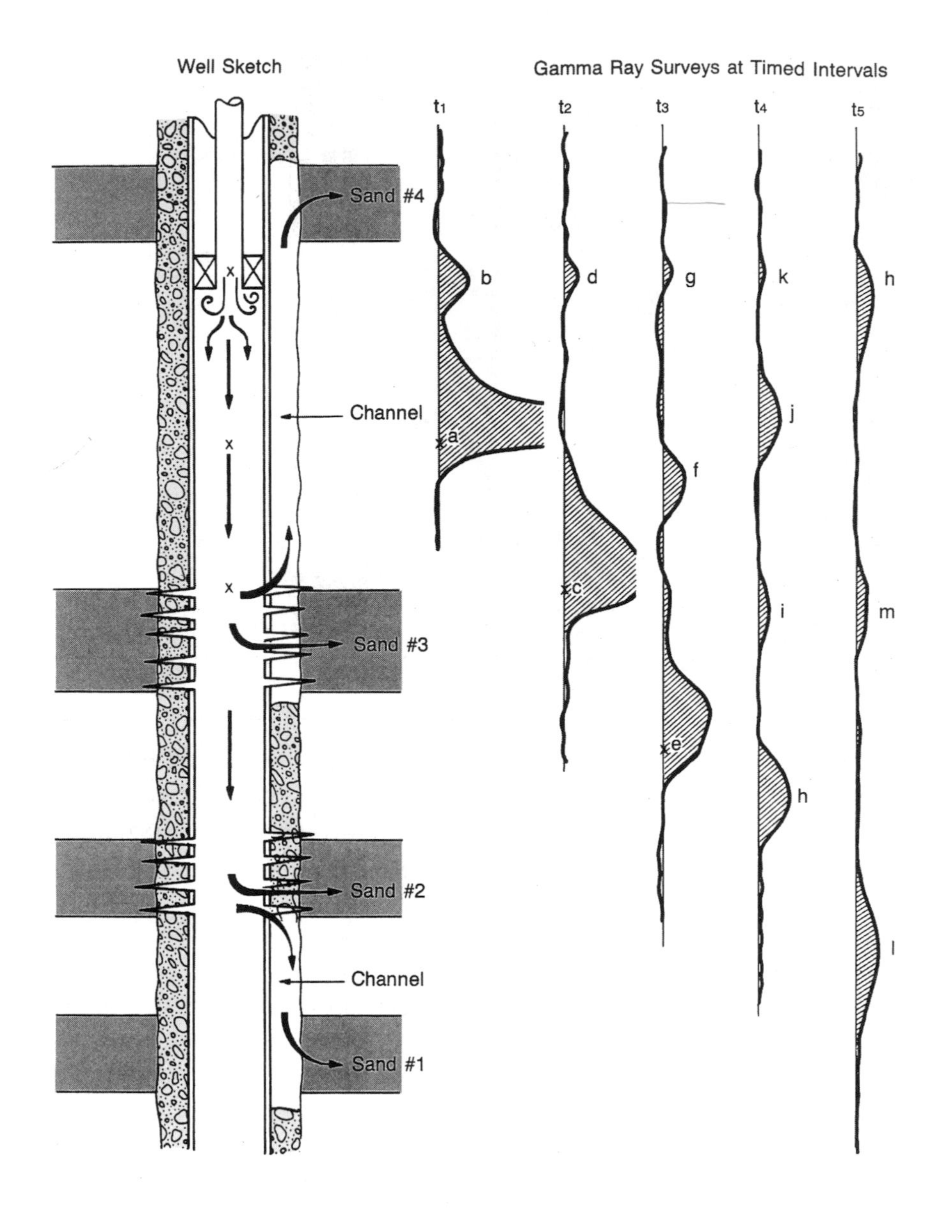

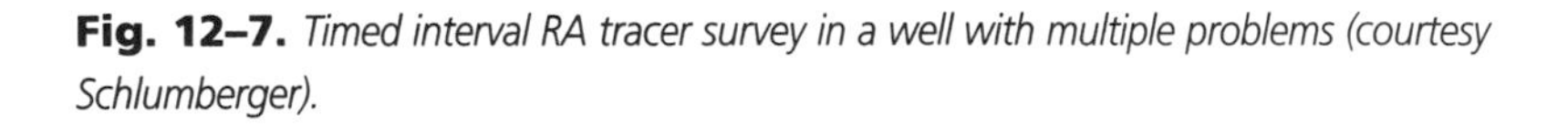
Fig. 12–7. *Timed interval RA tracer survey in a well with multiple problems (courtesy Schlumberger).*

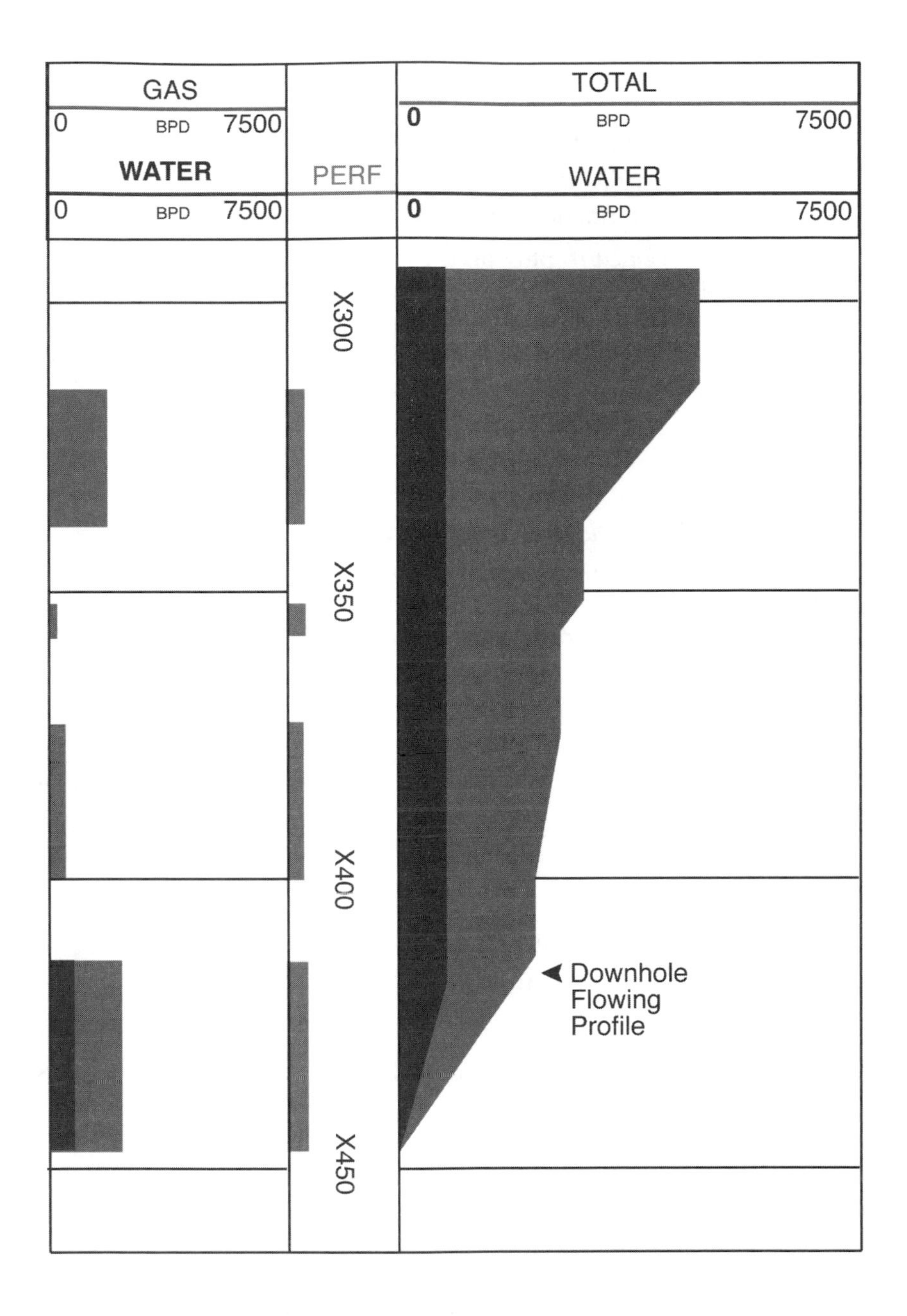

Fig 12–8. *Composite production log interpretation (courtesy Halliburton). Flow rates, fluid densities, holdups, and pressures are combined to provide a clear picture of downhole production. All of the water and a third of the gas were coming from the bottom set of perforations. These perforations were plugged off, and the water was eliminated.*

A full suite of production logs, including flowmeter, fluid density, pressure, and temperature measurements, was run with the well flowing. The figure indicates almost all of the water is coming from the bottom set of perforations. Further investigation with a pulsed neutron tool showed the water was coming from the formation and not from a channel.

The operator decided to plug off the bottom set of perforations with a through-tubing bridge plug. Water production was almost eliminated by the plugging operation. While total gas production was reduced by eliminating the bottom zone, flowing pressure was higher, allowing the upper zones to flow at higher rates. Also, the savings in operating costs by not having to dispose of the water resulted in a more profitable well.

Remedial Operations

Once the production logs have pinpointed and/or defined the problems, a solution and course of action must be decided on. Usually there are several options available to solve every production problem. The most versatile, and the most expensive, is to move in a workover rig, kill the well, pull the tubing, drill out the packer; squeeze off perforations, set plugs, or do whatever is thought necessary to solve the problems; then put the well back together again. Obviously, a complete workover is a very expensive option.

Other options entail working through the existing tubing with either coiled tubing or wireline units. These two options are often very successful and are much more cost-effective than a workover operation. Some of the operations that can be carried out on electric wireline are as follows:

- **Plugback operations**—new bridge plugs can be run through the tubing and set in the casing. Cement is often dumped on top of a plug to increase its strength.
- **Casing patch**—a sleeve is set inside the casing to seal off perforations or leaks.
- **Reperforate**—new zones can be opened or the number of shots in a zone can be increased.

CASING INSPECTION TOOLS

Casing and tubing are subject to several types of corrosion. Wells that produce hydrogen sulfide (H_2S) or carbon dioxide (CO_2) are particularly prone to corrosion because of the acids formed by these gases in the presence of water. This chemical attack is only one form that corrosion may take. Dissimilar metals that are immersed in an electrolyte (e.g., saltwater) experience electrochemical corrosion. Impurities, or even differences in composition within the metal, can create galvanic cells. Other corrosion mechanisms such as stress corrosion and hydrogen embrittlement may add to the attack on the casing. The longer a well has been in service, the more likely it is that some type of corrosion process is taking place. Corrosion may lead to casing leaks, burst pipe during a treatment involving pressuring the casing, or parted or collapsed casing.

One method of reducing or eliminating corrosion is with a cathodic protection system (Fig. 12–9). When galvanic corrosion takes place, a current loop occurs. Where the current leaves the casing, metal is eaten away; where the current returns to the casing, metal is added. The part of the well where corrosion occurs is the *anode*; the place where the current returns is the *cathode.* If an external electric current is applied to the casing with the opposite polarity or direction of current flow, corrosion is eliminated. The question, of course, is how much current. *Cathodic protection systems* work well but are expensive to install and operate.

Another anticorrosion scheme is *chemical inhibitors.* These are generally used in injection- or water-disposal wells. Inhibitors are added to the injected water on a continuing basis. The chemicals are expensive, and the amounts to use are determined by trial and error. Logs may be run routinely to monitor the presence or progress of corrosion taking place in a well and to help determine the effectiveness of the chemical treatments.

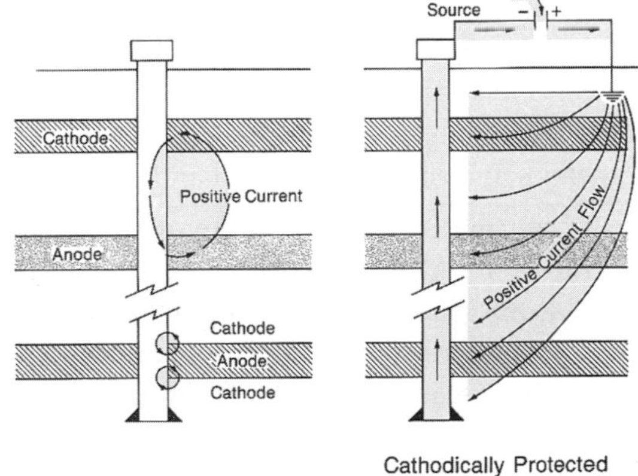

Fig 12–9. *Galvanic corrosion cell before and after cathodic protection was applied (courtesy Schlumberger).*

Casing Potential Tools

To determine whether galvanic corrosion cells are active in a well, a measurement of the naturally occurring electric potentials can be made. Casing potential tools measure the difference in potential between sets of electrodes in contact with the casing. The log shows which sections of the casing are being corroded by current flowing from the casing. The surveys can optimize the cathodic protection system so that no more reverse current flow is used than is needed to balance out the measured potentials.

Acoustic Imaging Tools

Today, most casing inspection logging is done acoustically with an acoustic imaging tool such as the CAST-V, UCI, or PET (see Chapter 11). The wavelength of the high-frequency signal emitted by the rotating transducer is much shorter than the thickness of the casing wall. The tools record ID, OD, and pipe thickness continuously over the logged interval. The log, as an image of the inside of the casing, provides diameter and thickness information; holes as small as 0.25 in. can be detected. The accuracy of the ID measurement is +/- 0.008 in.; the accuracy of the thickness measurement is about 2%.

The detail, the accuracy, and especially the ability to discern internal from external casing damage are this tool's strengths. The UCI log in Figure 12–10 shows how easily severe corrosion is detected—even a hole. One disadvantage of acoustic tools is they must have liquid in the wellbore so the sound energy can be transmitted into the casing and return.

Magnetic Flux Leakage Tool

The magnetic flux leakage tool has 8–12 pads, depending on casing size, that press against the pipe surface and measure magnetic flux anomalies due to changes in the casing quality. A centrally mounted electromagnet creates a low-frequency magnetic field in the casing. By using high-frequency eddy currents in addition to low-frequency signals, holes as small as 0.2 in. can be determined in gas- or liquid-filled environments. The tool is used to inspect pipelines as well as casing in oil and gas wells.

One interpretation problem with this tool is differentiating between corrosion and through-holes. In Figure 12–11, an area of corrosion is indicated from 3596–3530 ft. It is unclear whether the corrosion has completely penetrated the casing. Additional information, perhaps from an RA tracer survey, is required to verify a through-hole.

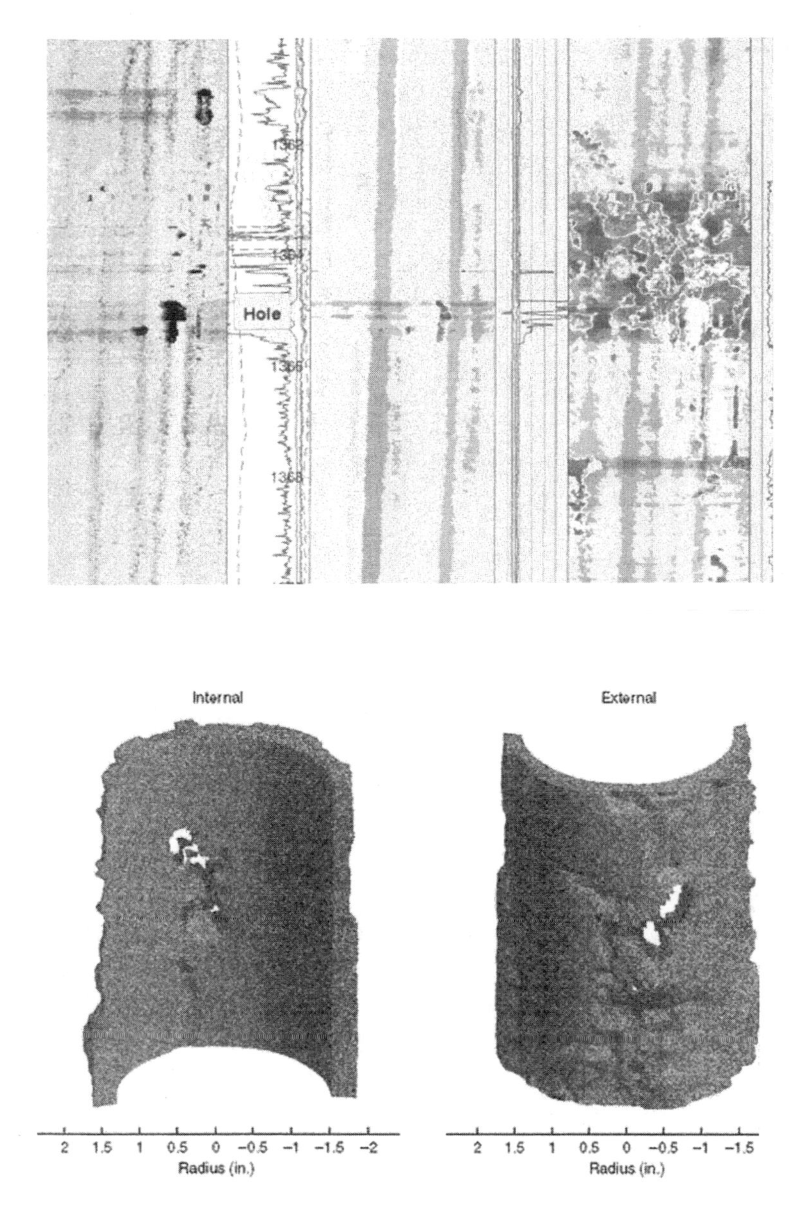

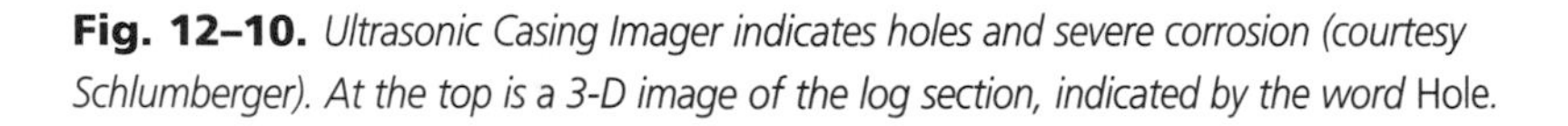
Fig. 12–10. *Ultrasonic Casing Imager indicates holes and severe corrosion (courtesy Schlumberger). At the top is a 3-D image of the log section, indicated by the word* Hole.

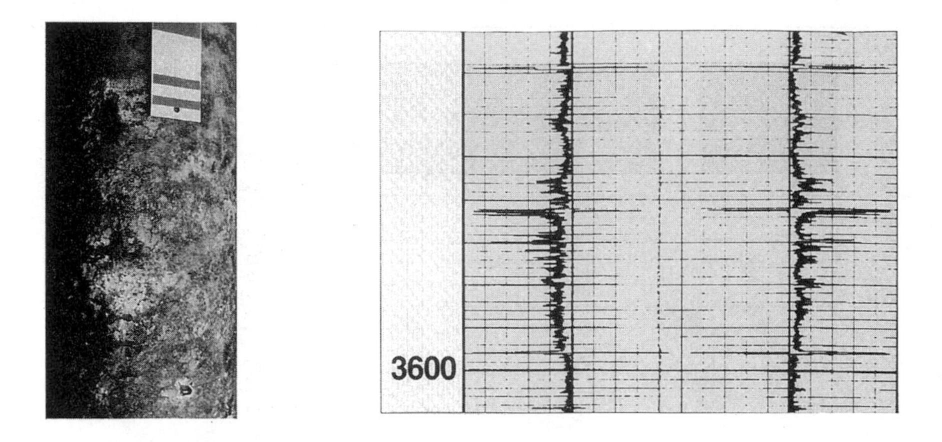

Fig. 12–11. *Flux leakage casing inspection log (courtesy Schlumberger). The curves indicate a casing problem at about 3582 ft.*

Multifinger Calipers

Multifinger calipers can be run to determine an accurate variation in ID. The devices have 16–80 fingers, depending on the casing ID. The caliper is sensitive to changes in diameter as small as 0.004 in. One advantage of this tool over the acoustic devices is that it can be run in any type of well fluid, even gas or air.

Remedial Operations

Severely corroded casing is difficult to repair, especially with wireline operations. A *casing patch* is sometimes effective if the corrosion occurs over a short interval. A *bridge plug* may be used to isolate a lower section of corrosion. Usually a workover rig will be necessary to remedy severe corrosion problems; sometimes the well must be plugged and abandoned.

THROUGH-CASING FORMATION EVALUATION TOOLS

The question might well be asked, "Why do we need cased-hole formation evaluation logs? The formation has already been evaluated by the openhole logs." Monitoring the reservoir, as it is producing or as it is undergoing secondary recovery operations, represents the greatest need for through-casing evaluation. By watching how and where the water saturation, porosity, and lithology change as the reservoir is produced, petroleum engineers can determine how best to maximize hydrocarbon recovery. Another reason to evaluate a formation through the casing is old wells: many may contain productive zones not recognized by the open-hole logs used when the well was originally drilled. Occasionally, wells don't even have open-hole logs because bad hole conditions existed.

The information obtainable from cased-hole logs is nearly the same as open-hole logs. Many of the open-hole tools also work in the cased-hole environment. However, the casing nearly always affects the measurement in some way. For example, on radiation measurements, the thickness of the steel in the casing reduces the count rate from the formation. This increases the statistical variation or uncertainty of the measurement.

Most through-casing evaluation tools depend on some type of neutron measurement, so we need to cover some nuclear physics before discussing the available measurements and the tools that make them.

Nuclear Physics and Through-Casing Tools

The highest energy neutrons used in logging today are about 14 MeV (14 million electron-volts). These are fast neutrons. A slow neutron has an energy level of about 1000 eV, epithermal neutrons are about 1 eV, and thermal neutrons are about 1/40 eV. The neutrons lose energy by elastic scattering, inelastic scattering, and absorption (also called *capture*) interactions with the elements present. The reactions decrease the energy of the neutrons until they disappear into their environment by diffusion.

Neutrons have *elastic reactions* initially (see Chapter 6 for a discussion of nuclear interactions). The atom that has the greatest effect on slowing the neutron is hydrogen because the mass of a hydrogen atom and a neutron are nearly the same. Since hydrogen atoms are found mainly in the formation fluids and the formation fluids are in the pore spaces, the rate at which the fast neutrons are slowed to lower energies is proportional to the porosity. Obviously, the liquid in the casing is also full of hydrogen atoms, but the effect of this signal is removed by the design of the tool circuitry.

Inelastic neutron scattering happens when a fast neutron strikes another atom and destabilizes, or excites, the atom. The neutron is slowed more than in an elastic interaction, but it still travels on, having other lower-energy interactions. The excited atom returns to a stable state by giving off an inelastic gamma ray. The gamma rays for each element have a characteristic spectrum of energy levels. By measuring the energy level versus count rates at several different times, the inelastic events are separated from other reactions such as the capture events. The carbon/oxygen (C/O) ratio is determined from the inelastic neutron scattering and is used to derive water saturation independent of formation water salinity. Since there is no carbon in water and no oxygen in oil or gas, the utility of the ratio is immediately obvious.

When the neutrons have decayed in energy to the thermal level, they are absorbed by the nucleus of an atom. The atom becomes excited and gives off a *capture gamma ray* to dissipate the extra energy. By measuring the spectrum of gamma-ray energies in the capture gamma-ray range, the amount of each of several different elements likely to be present can be measured. The yield, or amount, of silicon, calcium, chlorine, hydrogen, sulfur, and iron is determined by comparing the measured detector response to a library of responses and computing the most likely yield of each element.

Silicon is present in sand and shale, calcium is present in limestone and dolomite, chlorine is in formation water and shale bound water, hydrogen is in oil and water, and sulfur and iron are in shales and some other minerals. The yields of the different elements may be used to identify the lithology and to help refine the C/O ratio for water saturation.

A formation parameter called *macroscopic capture cross-section*, or *sigma* (Σ), can be measured by noting the rate at which the thermal neutrons are absorbed or captured by the formation. A capture gamma ray is released after an atom captures a thermal neutron. A plot of the total count rate of capture gamma rays versus time on a semilogarithmic scale shows an exponential decrease or decay rate. The decay rate is proportional to the sigma of the formation—a formation property such as density or resistivity. Sigma can be used to calculate water saturation if formation water salinity is known and is above about 35,000 ppm.

The natural gamma rays that occur in the formations are mainly due to uranium, thorium, and an isotope of potassium, K_{40}. The three elements are associated with certain minerals. By identifying the relative amounts of the three radioactive elements, a more accurate shale volume can be determined. In addition, the data can be combined with the spectra yield information for a more accurate lithology identification.

Nuclear measurements are possible only because of the development of newer, smaller, and better detector crystals with esoteric names like cerium-doped gadolinium oxyorthosilicate and bismuth germanate. Also, the computer is an intrinsic part of the new techniques. Without the computer to compare the tool response to a library of elements, and to calculate a best-fit solution, modern logging techniques would be impractical.

Now that we have an idea of how radiation tools respond to formations, let's discuss the tools available for through-casing formation measurements and what each can tell us. Several of the tools discussed earlier are used routinely in open-hole logging programs.

GR Spectrometry

The natural GR spectrometry tool as well as the older total GR tool can be used in cased-hole formation evaluation. As in open-hole evaluation, the GR measurement, is a shale indicator. The absolute value is reduced by the casing, and the statistical variation is increased due to lower count rates. The result is larger statistical error or uncertainty in the measurement. Nevertheless, the uranium, thorium, and potassium curves can be used along with other measurements to identify lithology.

CNL Log

The CNL tool normally is used to determine open-hole porosity. However, it works very well through casing. The computer compensates for the effect of the casing and cement. Older wells may have run early types of neutron logs, which are qualitative rather than quantitative in nature. In porosities below about 10%, the older neutron tool response is nearly linear; above 10%, the response is compressed. Casing effect and other environmental corrections are difficult to determine. Neutron tools are affected by shale and gas-bearing formations: shale increases the apparent neutron porosity, and gas reduces the apparent porosity.

Acoustic Tools

Acoustic velocity measurements can be made—porosity, synthetic seismograms, seismic check shots, and other seismic/acoustic measurements—if the cement is well bonded to both the casing and the formation. The acoustic (sonic) porosity combined with the neutron porosity log indicates gas zones because the sonic porosity is too high and the neutron porosity is too low in a gas-bearing formation. Porosity values must be corrected for shale, just as it is in open-hole interpretation.

Density Tools

Density tools can sometimes measure porosity through the casing, if the cement bonding is exceptionally good. The casing prevents photoelectric measurements, so only porosity is available, and it must be corrected as in open-hole interpretation. Naturally, density neutron crossplot porosity is ideal. Density neutron porosity comparisons can indicate gas-bearing formations since the neutron reads much too low and the density reads slightly high. By including acoustic porosity, lithology can be estimated. Unfortunately, conditions are seldom good enough for the density measurements to be valid.

Pulsed Neutron–Spectrometry Tools

The new generation of pulsed neutron–spectrometry measurements is the mainstay of cased-hole formation evaluation today. These tools, with their improved detectors and interpretation software, can determine water saturation from the C/O ratio independently of formation water salinity. They can also measure lithology and porosity from elemental yields of silicon, calcium, iron, sulfur, chlorine, and hydrogen. Formation sigma can be measured and used to determine water saturation when the formation water salinity is high enough.

Pulsed neutron tools are made in several different sizes. Older tools are about $3\frac{3}{8}$ in. in diameter and can be run in casing but are too big for tubing. The new tools are smaller and can be run through $2\frac{1}{2}$-in. and larger tubing. For 2-in. tubing, $1\frac{11}{16}$-in. C/O ratio tools with slightly reduced capabilities are available.

Pulsed Neutron–Decay Rate

Decay-rate tools measure sigma and a neutron porosity. Various gating and timing schemes are used to reduce statistics and to improve sigma measurements. Sigma can be used to calculate water saturation if the formation water salinity is high enough (about 35,000 ppm). While still used, these older tools ($1\frac{11}{16}$-in. to pass through 2-in. tubing) have been replaced largely by spectrometry measurement tools. In addition, pulsed neutron tools that measure only sigma and a neutron type porosity can be used in smaller tubing, if water salinity is high enough to make the sigma measurement definitive in determining water saturation.

The latest pulsed neutron tools can be run in combination with production logs on either flowing or injection wells. The tools can measure water flow inside or outside the casing by a process called *oxygen activation.* The neutron pulse activates the oxygen atoms in the water near the neutron generator. When the neutron generator ends the pulse, the slug of activated water flows past one or more detectors strategically located in the toolstring. The volume of water, or water holdup, flowing in the casing can be measured directly. The tool determines whether the water flow is inside the casing by shielding the individual detectors differently. Under favorable conditions, oil holdup can also be measured.

MONITORING THE RESERVOIR

As a field or reservoir is produced, some parameters change while others remain the same. Obviously, the amount of oil or gas in the reservoir usually decreases along with reservoir pressure; water saturation increases, and water production may also increase. If the field is in secondary or tertiary recovery, the original water salinity may have changed or may vary from place to place within the reservoir. Since permeability varies within zones, not all of the reservoir will produce at the same rate. The variation in permeability can lead to problems such as premature flood breakthrough and unproduced, or missed, hydrocarbons. When problems such as these are discovered, remedies can be attempted so that the reservoir can be produced in the most efficient and profitable manner possible.

Cased-hole formation evaluation logs monitor the reservoir and how it is producing. Sometime after a well is put on production, logs are run to create a base set of cased-hole parameters. Later monitor logs are compared with the base logs to determine how the well, or field, is behaving. Often, open-hole logging tools, such as the CNL and acoustic and density tools, if conditions permit, are run in the casing, and the log readings are corrected to agree with the open-hole logs.

Next, C/O ratios and elemental spectral yield logs are run, and the water saturations and lithology are brought into agreement with the open-hole logs. The C/O logs are then run routinely every few years to monitor the performance of the reservoir and/or flood. Usually a project such as this is undertaken on a large field where the results can be justified economically with increased production revenue or operational savings. Several wells in the field will usually be selected as monitor wells rather than monitoring every well in the field.

For an example of how reservoir-monitoring logs are used, look at Figure 12–12. The well is in a field undergoing waterflood. Water production in the field is very high because most of the oil has already been produced. A monitor log was run to determine if there were any bypassed oil zones that could be perforated to increase oil production. The well was shut in while a series of C/O logs were run over a period of several weeks.

The operator chose zone B based on open-hole logs and other information. The reservoir saturation tool (RST) indicated zone A as the most likely area to perforate to increase production; however, it also showed higher water and gas saturations in zone B than in zone A. So zone B was perforated and after a few weeks was stable at 200 bo/d with 95% water cut (4000 bw/d). Zone A was then perforated and commingled with zone B. Stabilized

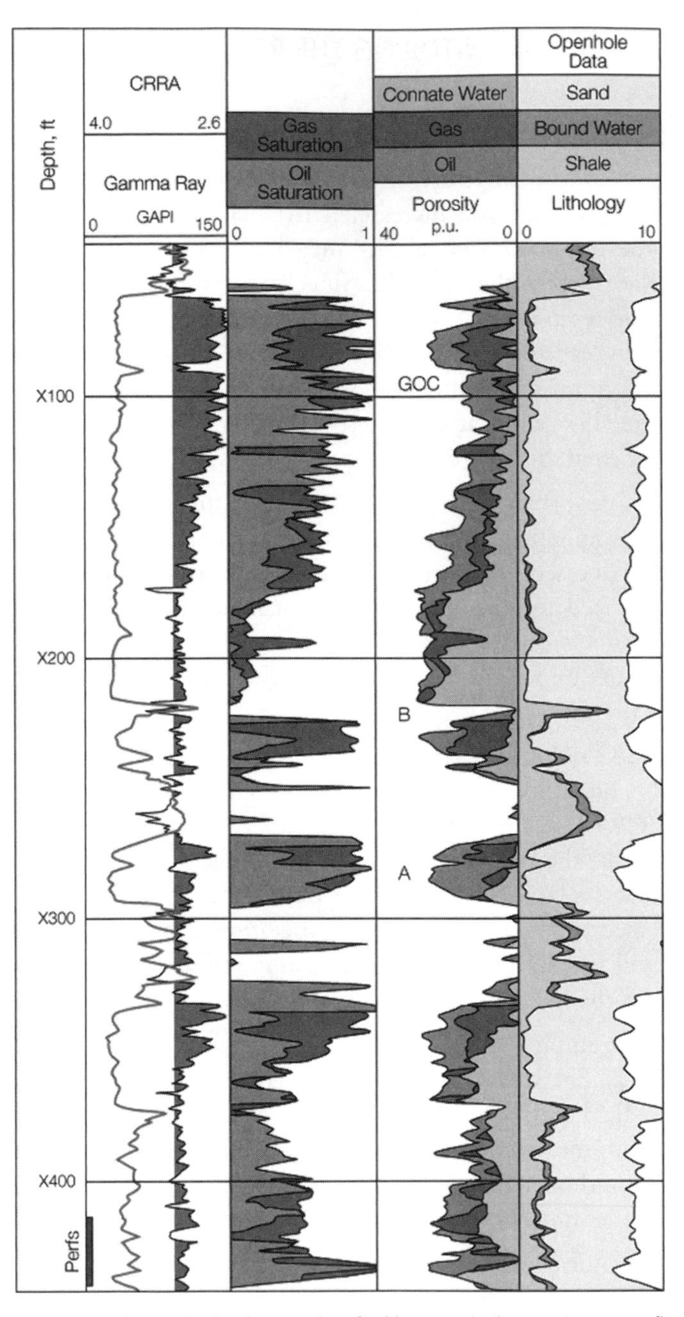

Fig. 12–12. *Reservoir monitoring log used to find bypassed oil zones in a waterflood (courtesy Schlumberger). Zone A was selected as the best prospect for recompletion.*

production increased to 600 bo/d with 90% water cut (6000 bw/d). Later production logs showed zone A was producing most of the oil and gas, and zone B was producing mainly water with some oil. Since this was a waterflood project, the large volume of produced water was treated and reinjected into the reservoir at an injection well. The monitor log was paid for in just a few days by the increased oil production.

Many oil fields have been producing for many years; during this span, production and reservoir engineering as a science was born and has matured. In the early days of oil production, little was known about how a reservoir behaved. Fields and reservoirs were severely damaged by unwise production practices, a lot of oil was left unproduced, and gas was flared or vented. Attempts were made to recover the missed oil by waterfloods, steamfloods, and firefloods, but again many mistakes were made during the learning process. Today, many of these older wells are candidates for through-casing formation evaluation. New reservoir monitor logs that can measure water saturation independently of formation water salinity and determine lithology and porosity are often able to find missed production. And prospecting for new production in old wells is often more successful, and much less expensive, than wildcat drilling.

A FINAL NOTE

Since the days of the gusher, memorialized in countless movies, there have been many changes in the oil industry. Not the least of these changes has been the birth and development of wireline logging. From a simple resistivity measurement powered by dry-cell batteries, measurements are routinely made today in wells more than 6 miles deep, with instrumentation as sophisticated as equipment sent to the moon or used in a modern hospital. The measurements still include resistivity, but they have expanded to include practically every physical parameter—acoustic velocity, electron density, response to powerful magnetic fields or to various types of radiation. As a better understanding of the reservoir brings the need for more and better information, the logging industry will continue to provide the tools and techniques to provide that information.

In Chapter 1 we asked the question, “Why run logs?” The remaining chapters gave the long answer to that question, but the short answer is: *for information.*

AFTERWORD

The logging industry has come a long way from that day in 1927 when two French brothers took some electrical measurements on an oil well in Pechelbronn, France. The early logging trucks had cable made of telephone wire wrapped in friction tape. The winches were turned by hand, and at every revolution of the winch drum, a bell rang. The bell signaled the helpers to stop turning the winch; then the engineer made a stationary measurement of voltage, which he converted to resistivity and recorded on a graph. These were the first logs.

From these very simple measurements, logging sophistication has advanced to space-age electronics, where spectroscopic measurements and downhole microprocessors are the norm. Logging trucks now carry enough steel-armored cable to log wells nearly 6 miles deep with bottom-hole temperatures in excess of 450 °F and pressures greater than 20,000 psi. Nearly 100 different measurements are available, from resistivity to depth to sonic waveforms to permeability.

The advent of the computer has caused a data explosion that is still growing. More ways of handling, presenting, and understanding the wealth of information available to us from logs are being developed daily. Where it will end is anybody's guess—perhaps with your computer calling a service company's computer and saying it has a well to log.

Since the first edition was written, a lot has happened in the oil patch. Although the oil companies' and the logging companies' computers are talking to each other, it still takes a human to make the important decisions. Ingenuity and perseverance in the face of grave difficulties, like $10/bbl oil, has once again paid off, and it looks like the oil industry has survived another bust. The hard times are good in a way, because they sort out the weak from the strong. Best of all, hard times promote innovation and new solutions to old problems. Today, we have a wealth of new and better logging tools that are able to work in much more difficult environments than just 10 years ago. There's no telling what the next 10 years will bring, but you can bet on it being better than what we have today. Meanwhile, as they say in Louisiana, *Laissez les bon temps roulet!* (Let the good times roll!)

SUGGESTED READING

For the reader who is interested in a more detailed discussion of logging tools, their application and interpretation, the following list of books and service company manuals should be helpful. In addition, many of the logging companies have Websites that contain the most recent information available.

Essentials of Modern Open-Hole Log Interpretation, John T. Dewan, Pennwell (1983)

Cased Hole and Production Log Evaluation, James J. Smolen, Pennwell (1996)

Manuals commonly available from the service companies:

Log Interpretation Manual
Includes theory, procedures, and applications

Log Interpretation Charts

Open Hole Services Catalog

Cased Hole Services Catalog

These manuals may be ordered on the internet

Baker Atlas
Website: *www.bakerhughes.com/bakeratlas*

Halliburton
Website: *www.halliburton.com*

Reeves Wireline
Website: *www.reeves-wireline.com*

Schlumberger
Website: *www.connect.slb.com*

Glossary

A

acoustic impedance. A function of the density and acoustic velocity of a material.

acoustic tool. *see* Sonic Log.

B

bond index. The percent of the casing circumference that is well bonded by cement.

bulk volume. The amount of substance present within a unit volume; abbreviated BV. Bulk volume is expressed as a percentage of the unit volume. If all individual bulk volumes are added together, they must equal 100%. (*see* Unit Volume)

bulk volume gas. The percentage of the unit volume that is gas; abbreviated BVG.

bulk volume hydrocarbon. The percentage of the unit volume that is hydrocarbon (includes both oil and gas); abbreviated BVH.

bulk volume matrix. The percentage of the unit volume that is non-porous rock; the mineral structure; abbreviated BVM.

bulk volume oil. The percentage of the unit volume that is oil; abbreviated BVO.

bulk volume water. The percentage of the unit volume that is formation water; abbreviated BVW.

bulk volume water—minimum. The BVW at which the formation water is bound to the formation grains by capillary forces; the same as BVW_{irr}. At BVW_{min}, no water will be produced from the formation.

C

capture. A type of reaction that occurs when a neutron is absorbed into or captured by an atom. The atom becomes highly energized and then releases the energy by emitting a gamma ray.

casing. The steel pipe that isolates the formations.

casing collar. A small piece of piping that connects two joints or pieces of casing together.

casing gun. *see* Perforating.

cement. A slurry-like material pumped into the space between the casing and the formation to isolate the formations from each other.

cement bond. The adherence of the cement to the casing. A good cement bond is necessary to have zone isolation.

centralizer. A device that physically positions a tool or the casing in the center of a wellbore.

check shots. A recording of one-way time versus depth, used to correlate acoustic logs to surface seismic data.

clastic. Rock formed from the fragments of other rocks.

compensated neutron log. Tool that measures in the same way as a neutron log but also compensates for hole rugosity, measures ratio of detector responses, and converts ratio to a linear porosity reading.

compressional wave. Type of sound wave that travels by compressing the material through which it travels; also called a primary wave or *P*-wave.

Compton scattering. Condition in which a gamma ray hits an electron and imparts some of its energy to that electron.

conglomerate. A poorly-sorted clastic sediment with large grains.

contamination gas. Gas introduced into the drilling fluid from a source other than the formation.

cores. Samples of the drilled formation that are obtained either by a core barrel on the drillpipe or by wireline tools that shoot a hollow bullet into the formation or use an electric motor and hollow bit to cut a core from the side of the wellbore.

correlation scale. Vertical scales of 1–2 in./100 ft. that geologists use to compare several wells over large intervals of formation.

cut. Leaching of oil from a sample by a solvent.

crossflow. Flow between two or more formations.

D

density. Weight of a unit volume of material divided by the weight of the same volume of water.

depth track. Column down the center of a log that records the depth of the well in multiples of 100 ft.

detail scale. Vertical scale of 5 in./100 ft. that can be used to detect more features than on a standard 1 or 2 in./100 ft. scale.

dielectric constant. A formation characteristic similar to density or acoustic velocity used to calculate water saturation without using formation water salinity.

dielectric permittivity. *see* Dielectric Constant.

dip. Angle a formation bedding plane makes with the horizontal. The direction the bed is dipping is called the strike.

drainage area. Area that a reservoir covers or that a well can drain.

driller's log. Record of what occurs on a drilling rig, recorded by depth; notes types of rock encountered, rate of drilling, oil or gas flows, equipment breakdowns, accidents, and any other occurrence that might have a bearing on evaluating the well.

drilling rate curve. Rate of penetration expressed in units of length per hour.

drilling time curve. Rate of penetration expressed in minutes per unit of length.

drillstem test. Technique that uses an open-hole packer to isolate the zone of interest from the hydrostatic pressure of the mud column. If the zone is productive, formation fluids will flow into the drillpipe. Valves control the flow and dispose of the fluids. A pressure recording is made of all events and is later analyzed.

dual induction laterolog. Tool developed for areas of low to moderate resistivity and deep invasion; has two induction curves (deep and medium) plus a shallow-reading laterolog curve. (*see* Induction Log)

dual laterolog. Tool developed for areas of high resistivity and deep invasion; has deep and medium laterolog curves. (*see* Laterolog)

effective porosity. Pore space available to hold fluids.

E

electric log. Tool that emits current from a constant electrode source and then measures the current at another electrode some distance away with respect to a reference electrode. The earliest form of resistivity log.

electric wireline. Wire rope with insulated electrical wires or conductors beneath the strands of cable.

epithermal neutron log. Tool that uses lower-energy neutrons to measure porosity, especially in air-filled holes.

evaporates. Rocks formed from precipitate residue after a salty body of water evaporates.

F

flushed zone. Area where filtrate has flushed out all original fluids possible; abbreviated *xo*.

focused electric log. Tool used on highly resistive formations; the measuring current is forced into the formation by guard electrodes.

fracture porosity. Porosity attributed to fracture planes rather than to pores; the porosity is low, but the permeability is high.

free fluid index. Percentage of the pore space that is filled with movable fluids.

full-bore spinner. A tool that uses a collapsible centralizer and propeller flowmeter to go through tubing and then extends itself to measure fluid flow in the casing.

G

gamma-ray log. Tool that reads gamma rays emitted naturally from formations.

gas chromatography. Method of analyzing the composition of the gas stream on a regular but intermittent basis.

geophone. An acoustic transducer used to record seismic data.

ground loops. Circular currents generated by an induction tool that are concentric with the tool's axis.

H

header. The top of the log, where well data are noted.

holdup. The percentage of each phase of fluid in a multiphase fluid.

holdup meter. A device that measures holdup; may be mechanical, electrical, optical, or nuclear.

hydrostatic pressure. Pressure exerted by a column of fluids usually associated with the wellbore.

I-J

imaging logs. Tools that take a multitude of measurements radially. Values are converted to color and combined into a virtual image of the wellbore.

induction electric log. Tool with a single induction curve combined with a short normal or shallow laterolog curve, used in medium- to high-porosity formations.

induction tool. Device that uses an alternating magnetic field to create ground loop currents in the formation concentric with the well-bore. Formation resistivity is inversely proportional to the amount of current induced in the formation.

intergranular porosity. *see* Matrix Porosity.

invaded zone. Area including the flushed zone, and extending into the formation to the depth that the wellbore fluids have mixed with the formation fluids; denoted as subscript $_i$.

invasion. A condition in which mud filtrate penetrates the formation next to the wellbore. The mud filtrate may cause swelling of the shale in the formation, which restricts production.

irreducible water saturation. The minimum possible water saturation in a formation; a function of grain size, surface area of the sand grains, and shaliness. A zone at irreducible water saturation produces all hydrocarbons and no water.

K

kick. A condition in which formation fluids flow into a well without being controlled. A major kick is called a blowout.

L

lag. The amount of time that elapses from the moment when the drill bit penetrates a new formation until the moment when the downhole particles and/or traces of gas circulate to the surface.

lateral. A specific arrangement of electrodes on an electric log. The lateral curve reads more deeply into the formation than a normal curve.

laterolog. A simple focused log with electrodes that force any current into the formation to minimize the effect of the low-resistivity borehole.

liberated gas. Fluid released from exposed pores in a formation that mixes with the drilling mud and flows back to the surface.

lithology. Rock type, such as sandstone, shale, limestone, dolomite, anhydrite.

log. Data recorded versus depth or time in graph form or with accompanying written notes. A common term for any tool run downhole that generates or measures signals recorded on the printed log.

long normal. *see* Normal.

lost circulation. A condition in which large amounts of drilling mud are pumped into the formation and lost in fractures or vugs.

M

magnetic resonance imaging. Also called nuclear magnetic resonance. Measures the free fluid index, water saturation independent of formation water salinity, and permeability.

matrix. The mineral structure from which a formation is made.

matrix porosity. Pore spaces between rock grains whose nonrock volume equals the porosity.

microlaterolog. A smaller version of a laterolog, used in high-resistivity formations; measures flushed-zone resistivity. Abbreviated MLL.

microlog. An early resistivity tool with a printout consisting of a caliper curve and two resistivity curves; most effective in low- to medium-resistivity formations. Used to indicate permeability.

microresistivity logs. Tools designed to read the resistivity of the flushed zone; a very shallow-reading tool.

microspherically focused log. A pad device whose readings are recorded on a logarithmic scale. Useful in helping determine depth of invasion, moved hydrocarbons, permeability, porosity, hole diameter, and zone thickness; abbreviated SFML

mud cake. Sealant present in the drilling mud used to protect the formation from invasion.

N

neutron log. Tool that bombards formations with neutrons from a radioactive source housed in the tool, and that measures the porosity as a function of the number of hydrogen atoms present.

normal. A specific electrode arrangement on an electric log; often categorized as short normal or long normal, depending on spacing.

O

offset VSP. A VSP made with the sound source removed some distance from the wellsite. (*see* VSP)

offset well. A nearby well whose logging data and production history are assessed during plans to drill a new well in an existing field.

Ohm's law. A principle that states voltage is equal to current multiplied by resistance.

ohms. In logging terminology, an abbreviation for ohm-meters (meters squared per meter). In the field, it is referred to as "ohms," although purists prefer the longer version.

oil saturation. The percentage of the pore space filled with oil; abbreviated *S*. If no gas is present, $S_w + S_o = 1$.

P

packer flowmeter. Tool that uses an umbrella-like element to divert the well fluids through the measuring section of the tool. Used mainly in lower-flow-rate wells. Also called a diverter flowmeter.

perforations. Holes shot in the casing so well fluids can flow to the surface. The perforations are usually made by guns using shaped charges. Occasionally, bullets are used.

perforating gun. Tool used to shoot holes in the casing. Bullets or, more normally, shaped charges make the holes. The guns can be run on wireline or on the tubing.

production logs. Tools run on producing wells to determine the flow profile and to identify problems.

photoelectric effect. Condition that occurs when a low-energy gamma ray, passing close to an atom, is absorbed; an electron is then ejected into space.

porosity cutoff. The minimum porosity established for production. Formations with lower porosity are not counted as net pay.

pressure differential. The difference between the wellbore or hydrostatic pressure and the formation pressure.

primary wave. *see* Compressional Wave.

pulsed neutron logs. Carbon–oxygen logs, thermal decay time logs, and induced spectrographic logs. The logs use a pulsed neutron source but different neutron energy levels to evaluate reservoir parameters through the casing.

R

rate of penetration. Speed at which a bit penetrates a formation; abbreviated ROP.

recovery factor. Percentage of oil or gas recoverable by primary production; used in the reserve calculation.

refraction. The change in the speed of sound waves when the material through which they travel changes.

regional dips. General dip of the formations over a large area.

reserves. Amount of recoverable oil or gas in a formation.

residual oil. Oil that remains in the formation and that cannot be removed by normal means; residual oil saturation is abbreviated S_{or}.

resistance. Resistivity times length through which current flows divided by the cross-sectional area; $r = pL/A$.

resistivity. A physical parameter or property of a material. A measure of the difficulty an electric current will encounter in passing through the material, measured in ohm-meters.

resistivity profile. Difference in readings between shallow, medium, and deep resistivity curves.

rugosity. Roughness.

S

secondary wave. *see* Shear Wave.

seismic. Energy transmitted through the earth. Surface seismic is recorded using long lines of geophones and an energy source. The recording is called a seismogram. Seismic data can reveal structures or traps in the earth's formations.

shale baseline. A fairly constant reading opposite shales, used as a guide on the spontaneous potential log.

shale bound water volume. Water chemically bound to the shale and not free to move.

shaped charge. An explosive device used to make holes in the casing.

shear wave. A sound wave slower than a compressional wave that moves vertically rather than horizontally to the axis of the wave; also called an S-wave or secondary wave.

short normal. *see* Normal.

sidewall neutron porosity tool. A device with skid-mounted detectors pressed against the borehole wall, generally in air-drilled holes; abbreviated SN.

slip velocity. Difference in velocity between two fluid phases.

solution gas. Gas dissolved in the oil, at reservoir conditions of temperature and pressure.

sonde. Synonym for logging tool.

sonic log. Tool that uses sound waves to measure porosity in a formation.

spontaneous potential. A naturally occurring voltage caused when conductive drilling mud contacts the formations; abbreviated SP.

stacking. The practice of connecting logging tools so they can be run simultaneously.

static SP. The maximum spontaneous potential that can be measured if no current flows in the borehole; a constant reading; abbreviated SSP.

stratigraphic dip. The dip associated with the deposition of a feature, such as a point bar or a braided stream.

structural dip. The dip associated with a structure, such as a fault or a salt dome.

stroke counter. A mechanical device used to count the strokes of reciprocating mud pumps. Since the volume of mud the pump moves with each stroke is known, the rate and volume of mud pumped can be determined by counting the strokes. The hole volume and the displacement of the drilling string can be estimated accurately. The number of strokes on the stroke counter necessary to circulate the cuttings to the surface can be calculated from this information.

synthetic seismogram. A record made using formation density and acoustic velocity from the well logs. The result is used to refine the interpretation of the surface seismic.

T

tadpole plot. A presentation of the dipmeter information. The swarm of data and the directional "tails" resemble tadpoles to some peoples' eyes.

thermal catalytic combustion. Also known as hot-wire detection (HWD), one of the main ways to detect gas in the drilling mud; abbreviated TCC.

three-phase flow. Exists when gas, oil, and water flow simultaneously.

through-tubing perforating. Technique that uses a shaped-charge gun small enough to fit through the tubing.

total porosity. Percentage of nonrock volume in a rock; may be filled with oil, gas, formation water (including bound water), or shale. Porosity is often referred to as void space in the rock—a misnomer.

track. Vertical column, often numbered 1, 2, or 3, in which a specific reading from a logging tool is recorded.

transmitter current. High-frequency current generated by an induction tool.

true resistivity. Resistivity of an uninvaded or uncontaminated zone; abbreviated R_t.

U

unit volume. A cube of formation one unit long on each side. The volume (V) of a unit volume is 1 unit × 1 unit × 1 unit = 1 unit.

V

vertical resolution. The thinnest bed a logging tool will detect.

vertical seismic profile. Recording a seismogram with the geophones located in the well bore at several depths; abbreviated VSP. Usually the sound source is located near the wellsite. In special cases, the sound source may be moved in a particular direction or placed in several different locations. (*see* Walkaway VSP and Offset VSP.)

virgin zone. Undisturbed or uncontaminated formation.

Vug. A space in the rock that is larger than a pore, often ranging up to cavern size.

vugular porosity. Porosity attributed to caverns or vugs in the rock, usually caused by dissolution of the rock by water movement.

W–Z

walkaway VSP. A record made by moving the sound source farther and farther from the wellsite while recording a vertical seismic profile (VSP). Often used to locate salt domes.

water saturation. Percentage of the pore space filled with formation water. If a formation's porosity is completely filled with water, S_w = 100%. If oil and water are present, $S_w + S_o = 100\%$.

weighting material. Any solid (such as barite) added to the drilling mud to raise the fluid pressure of the mud column.

zone isolation. Condition that exists when there is no flow between formations behind the casing. The cement is well bonded to both the formation and the casing.

NOMENCLATURE

SUBSCRIPTS

Traditional Subscripts	Standard SPE of AIME and SPWLA	Standard Computer Subscripts	Description	Example
a	a	A	apparent (general)	R_a
b	b	B	bulk	ρ_b
bh	bh	BH	bottom hole	T_{bh}
cor,c	cor	COR	corrected	t_{corr}
cp	cp	CP	compaction	B_{cp}
dol	dol	DL	dolomite	t_{dol}
e,eq	eq	EV	equivalent	R_{weq}, R_m
f,fluid	f	F	fluid	ρ_f
fm	f	F	formation (rock)	T_f
g, gas	g	G	gas	S_g
gxo	gxo	GXO	gas in flushed zone	S_{gxo}
h	h	H	hole	d_h
h	h	H	hydrocarbon	ρ_h
hr	hr	HR	residual hydrocarbon	S_{hr}
i	i	I	invaded zone	d_i
irr	i	IR	irreducible	R_i
J	j	J	liquid junction	E_j
log	LOG	LOG	log values	t_{LOG}
ls	ls	LS	limestone	T_{ls}
m	m	M	mud	R_m
max	max	MX	maximum	$\varnothing_{max}$
ma	ma	MA	matrix	t_{ma}
mc	mc	MC	mud cake	R_{mc}
mf	mf	MF	mud	R_{mf}
mfa	mfa	MFA	filtrate	R_{mfa}
min	min	MN	mud filtrate, apparent	
o	o	O	minimum value oil (except with resistivity)	S_o
or	or	OR	residual oil	S
o	o	ZR	with resistivity– 100% water saturated	R_o
r	r	R	relative	k_{ro}, k_{rw}
r	r	R	residual	S_{or}, S_{hr}

Traditional Subscripts	Standard SPE of AIME and SPWLA	Standard Computer Subscripts	Description	Example
sd	sd	SD	sand	
ss	SS	SS	sandstone	
sh	sh	SH	shale	V_{sh}
SP	SP	SP	spontaneous potential	E_{SP}
SSP	SSP	SSP	static SP	E_{SSP}
t	t	T	true (as opposed to apparent)	R_t
T	t	T	total	C_t
w	w	W	water, formation water	S_w
xo	xo	XO	flushed zone	R_{xo}
IL	I	I	from Induction Log	R_t
Ild	ID	ID	from Deep Induction Log	R_{ID}
Ilm	IM	IM	from Medium Induction Log	R_{IM}
LL	LL (also LL3, LL8, etc)	LL	from Laterolog (also Laterolog 3, etc)	R_{LL}
6FF40			from 6FF40 induction log	R_{6FF40}
MLL	MLL	MLL	from Microlaterolog	R_{MLL}
PL	P	P	from Proximity log	R_P
N	N	N	from normal resistivity log	R_N
D	D	D	from Density Log	$Ø_D$
N	N	N	from Neutron Log	$Ø_N$
SNP	SN	SN	from Sidewall Neutron Log	$Ø_{SN}$, $Ø_{SND}$
CNL	CN	CN	from Compensated Neutron Log	$Ø_{CN}$
TDT	PNC	PNC	from Pulsed Neutron Log	$Ø_{PNC}$, $Ø_{TDT}$
S	SV	SV	from Sonic Log	$Ø_{SV}$
ND			from combined Neutron and Density Log	O_{ND}
GR	GR	GR	from Gamma Ray Log	$Ø_{GR}$

References: *Supplement V to 1965 Standard–"Letter and Computer Symbols for Well Logging and Formation Evaluation," in* **Journal of Petroleum Technology** *(October 1975), pages 1244–1261, and in* **The Log Analyst** *(November/December 1975), pages 46–59.*

SYMBOLS

Traditional Symbols	Standard SPE of AIME and SPWLA	Standard Computer Subscripts	Description	Customary Units or Relationship
a	K_R	COER	coefficient in F_R, Ørelation	$F_R=K_R/\varnothing^m$
c	C	ECN	conductivity, electric	millimhos per meter (mmhos/m)
Cp	B_{ep}	CORP	sonic compaction correction factor	
D	D	DPH	depth	
d	d	DIA	diameter	
F	FR	FACHR	formation resistivity factor	
H	I_H	HYX	hydrogen index	
h	h	THK	thickness (bed, mudcake, etc.)	
K	K_e	COEC	electrochemical SP coefficient	
k	k	PRM	permeability, absolute (fluid flow)	
m	m	MXP	porosity cementation exponent	
n	n	SXP	saturation exponent	
p	p	PRS	pressure	
R	R	RES	resistivity	
r	r	RAD	radial distance from hole axis	
S	S	SAT	saturation	
T	T	TEM	temperature	
BHT, T_{bh}	T_{bh}	TEMBH	bottom-hole temperature	
FT, T_{fm}	T_f	TEMF	formation temperature	
t	t	TIM	time	
v	v	VAC	velocity (acoustic)	
V	V	VOL	volume, volume fraction	
ϕ	ϕ	POR	porosity	
Δt	t	TAC	sonic interval traveltime	
ρ	ρ	DEN	density	
Σ	Σ	XST	neutron capture cross section	

INDEX

A

B

C

D

E

F

G

H

I

J

K

L

M

N

O

P

Q

R

S

T

U

V

W

Z